EINSTEIN AT HOME

ALSO BY JOSEF EISINGER

Einstein on the Road

EINSTEIN AT HOME

FRIEDRICH HERNECK

Translated, with an Introduction by

JOSEF EISINGER

Foreword by ALICE CALAPRICE

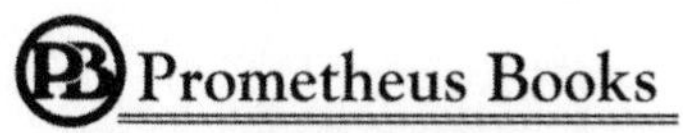
59 John Glenn Drive
Amherst, New York 14228

Published 2016 by Prometheus Books

Originally published as *Einstein privat* by Buchverlag Der Morgen, Berlin, 1978.

This translated and expanded edition of Friedrich Herneck's *Einstein privat*
has been made possible through the kind cooperation of Mr. Uwe V. Lobeck,
Herneck's literary executor.

Cover design by Grace Conti-Zilsberger
Cover image © Bettmann/CORBIS

Inquiries should be addressed to
Prometheus Books
59 John Glenn Drive
Amherst, New York 14228
VOICE: 716–691–0133
FAX: 716–691–0137
WWW.PROMETHEUSBOOKS.COM

20 19 18 17 16 5 4 3 2 1

Library of Congress Cataloging-in-Publication Data Pending

ISBN 978-1-63388-146-4 (pbk)
ISBN 978-1-63388-147-1 (ebook)

Printed in the United States of America

In my student years, during the Second World War, it was my good fortune to be a member of the Mendel family household in Toronto. Bruno and Hertha Mendel, but particularly Hertha's mother, Toni Mendel, had been close to Einstein in Berlin, in the years before Hitler. At that time, they, along with Bruno and Hertha's three children, had all lived in Toni Mendel's copious Wannsee villa, where Einstein was a frequent visitor. As a result, I first "got to know" Einstein through the eyes of the Mendels, and I dedicate this little book to the memory of these bighearted and cultured friends of long ago,

Toni, Bruno, and Hertha Mendel.

J. E., New York, August 2015.

CONTENTS

FOREWORD

In the days before reality TV, people rarely had the opportunity to encounter information about the day-to-day homelife of celebrities they either admired or despised. In the late 1970s, German historian of science Friedrich Herneck decided to change all that in the case of Albert Einstein, the most celebrated and revered scientist of our time.

In 1978, Herneck interviewed Herta Waldow, who, before her marriage, was known as Herta Schiefelbein. At the young age of twenty-one, in 1927, she became Albert and Elsa Einstein's live-in housekeeper in Berlin, staying in their apartment until June 1933. These chaotic years turned out to be the final ones the Einstein couple spent in Germany.

Fräulein Schiefelbein, who was several years younger than Einstein's stepdaughters, Ilse and Margot, became a well-respected and beloved member of the Einstein household. She had her own room while living in the Einsteins' spacious apartment on Haberlandstrasse, located in today's Charlottenburg-Wilmersdorf district in the fashionable southwestern part of the city's center. Ilse was already married by the start of Herta's employment and had left the apartment, but Margot continued to live there until her own marriage in 1930. Both women and their husbands stayed in the apartment occasionally on visits, and Margot and her husband apparently lived there sometime during 1933. Fräulein Schiefelbein was also with the Einsteins when they stayed in their summer home in Caputh, where they entertained many visitors and their children.

In early 1933, after the Einsteins returned to Europe following their stay at Caltech in California, they did not go back to Berlin. For safety's sake, they moved to Belgium for a few months, as Hitler tightened his grip on Germany and harassment of Jews intensified. They trusted

Fräulein Schiefelbein to remain in the Haberlandstrasse apartment, and she was there when many of the Einsteins' belongings were confiscated by the Nazis that spring. She also helped Ilse and Margot pack up or dispose of the remaining items. Albert and Elsa Einstein remained in Belgium, and he for a short time in England, until their emigration to America in the fall. Einstein never again returned to Europe.

Herta Schiefelbein was not a gossipy sort but could hardly avoid being the proverbial fly on the wall during her life with the Einsteins. A keen, perceptive, and informed observer of the daily goings-on in the fifth-floor apartment and in Caputh, she appears to have fitted well into the household and sustained a respectful confidentiality about the things she heard and saw. In the five interviews presented in this book, she gives straightforward accounts of daily life at home; Einstein's political and scientific activities and visitors; his habits, friendships, and liaisons; and the final days of the Einsteins' time in Berlin. Herneck, in turn, does not ask questions of a suggestive nature and keeps the interviews at a high level.

Prior to the publication of Herneck's interviews with Frau Waldow in what was then East Germany, biographers have had little to work with in reconstructing the day-to-day routines of Einstein's domestic life. Even then, this information was restricted to those who were able to read and understand German, though some biographers writing in English have recounted some of Frau Waldow's recollections in their own books. Now, however, we have the complete interviews in a lively and reliable English translation. Josef Eisinger's own notes add texture and context to the conversations.

In 1953, about a year and a half before his death, Einstein wrote to his biographer Carl Seelig: "In the past, it never occurred to me that every casual remark of mine would be snatched up and recorded. Otherwise I would have crept further into my shell." Einstein's grievance is understandable, and he was never able to come to terms with the price of fame. Perhaps he would have objected to the publication of a book like this, but dozens of others, on a variety of topics, have, over the years, already "snatched up and recorded" not only his remarks but also his involvements with the variety of organizations and individuals mentioned in this book.

This kind of personal information has been valuable in giving us a full portrait of the man, allowing us to abandon the one-dimensional caricature of an absent-minded professor who did nothing but think in equations and conduct thought experiments. This book shows us that, in many ways, the multifaceted physicist was just as human and vulnerable as the rest of us.

Alice Calaprice, Claremont, California, 2015.

ACKNOWLEDGMENTS

I am grateful to Mr. Uwe V. Lobeck, the Director of the Friedrich Herneck Archive in Dresden, for supporting this project and for giving permission to publish this translation of Herneck's book *Einstein privat*. I thank him also for making some of its illustrations available for the present volume.

I am indebted to Professor Dieter B. Herrmann for contributing the essay about Friedrich Herneck that appears in translation as chapter 2. I am also grateful to him for helping me make contact with the Friedrich Herneck Archive and for his unremitting encouragement and advice.

I am grateful to my friend and experienced Einstein biographer Alice Calaprice for contributing a foreword and for her unstinting advice. My thanks also to Barbara Wolff of the Einstein Archive at the Hebrew University, Jerusalem, for her kind cooperation. I am indebted to my editor, Steven L. Mitchell, for several helpful discussions and suggestions, and to my editor, Sheila Stewart, for her intelligent and proficient copyediting. Finally, my special thanks to Dr. Frank Mecklenburg of the Leo Baeck Institute, New York, for his cooperation and his benevolent support.

Many thanks, also, to Lesley Wyle for her archival research, to Alison Eisinger for copyediting portions of this book, and to Simon Eisinger, who sifted through the available literary, photographic, and architectural evidence in order to reconstruct the likely layout of Einstein's Haberlandstrasse apartment. My warm thanks, finally, to my wife, Styra Avins, for being such a good mate for more than fifty years.

J. E., New York, 2015.

TIME LINE

1876—Pauline Koch and Hermann Einstein are married in Ulm.

1879—Albert Einstein is born on March 14 in Ulm.

1880—The Einsteins move to Munich to found an electrical manufacturing company.

1880—Albert's sister, Maria (Maja), is born in Munich.

1885—Albert enters school. He receives violin lessons—for a short time.

1889—Einstein is introduced to science and geometry books. He enters Luitpold Gymnasium.

1894—Einstein's parents move to Italy. Albert, left behind, joins family in December.

1895—Einstein fails to enroll in Federal Polytechnical School (Polytech), enters school in Aarau.

1896—Einstein enters Polytech, meets Mileva Marić. He relinquishes German citizenship.

1900—Einstein graduates from Polytech, begins futile search for employment.

1901—Einstein's first physics paper published. He tutors privately.

1902—Begins work at Patent Office in June. Hermann Einstein dies.

1903—Einstein marries Mileva in January.

1904—Einstein and Mileva's first son, Hans Albert, is born.

1905—*Annus mirabilis*. Einstein publishes seminal papers in relativity and quantum physics.

1909—Associate professorship at the University of Zurich

1910—Einstein and Mileva's second son, Eduard (Tetel) is born. Maja Einstein marries Paul Winteler.

1911—Professorship at the German University of Prague.

1912—Professorship at the ETH, Zurich.

1913—Planck and Nernst offer Einstein membership in Prussian Academy and professorship at the University of Berlin. He accepts.

1914—WWI begins. The Einstein family moves to Berlin; their marriage fails; Mileva and the boys return to Zurich.

1915—General relativity theory published.

1918—WWI ends. Divorce agreement reached with Mileva.

1919—Marriage to Elsa. Eddington expedition confirms general relativity theory. World fame ensues.

1920—Pauline Einstein, Albert's mother, dies. Anti-relativity rally in the Philharmonie.

1921—First America trip with Chaim Weizmann.

1922—Einstein and Elsa leave for Japan lecture tour. They will visit Ceylon, Shanghai, Palestine, Spain, etc. Nobel Prize awarded to Einstein for photoelectric effect.

1924—Elsa's eldest daughter, Ilse, marries Rudolf Kayser.

1925—Lecture tour in Argentina, Uruguay, and Brazil.

1927—Hans Albert marries Frieda Knecht. Einstein and Niels Bohr begin dispute on quantum mechanics. Herta Schiefelbein joins Einstein household.

1928—Einstein ill with heart ailment. Helen Dukas hired as secretary.

1929—Einstein turns fifty. Summer house in Caputh is built. *Tümmler*, Einstein's sailboat, is delivered.

1930—Elsa's younger daughter, Margot, marries Dmitri Marianoff. Einstein and Elsa leave for the first winter session (January to March 1931) at Caltech, in Pasadena, California.

1933—Third winter session in Pasadena. Hitler attains power in Germany, January 30. After six months in Belgium, Einstein and Elsa settle in Princeton.

1934—Ilse critically ill in Paris; Elsa rushes to her side.

1935 The Einsteins acquire their final home at 112 Mercer Street, Princeton, New Jersey.

1936—Death of Elsa in Princeton.

1939—WWII starts. Maja arrives in Princeton. Atomic weapons letter to Roosevelt.

1940—Einstein, Margot, and Helen Dukas become US citizens.

1945—Hiroshima bomb. WWII ends. Einstein retires from the Institute of Advanced Study.

1948—Death of Mileva in Zurich.

1951—Death of Maja in Princeton.

1955—Death of Einstein in Princeton.

1986—Death of Margot in Princeton

CHAPTER 1
INTRODUCTORY REMARKS

Elsa Einstein, Albert's wife, must have been in an uneasy frame of mind as she headed for the employment agency in Berlin's Jägerstrasse, that morning in June of 1927. She had had little luck with domestic servants, but the departure of her latest housekeeper had been the last straw: Not only had the woman left her job without giving notice but she had taken several pieces of the family silver with her. For the wife of a world-famous scientist who is obliged to host many dinner parties and other social gatherings, it loomed as a calamity. But lo, this was to be Frau Elsa's lucky day! At the agency she met the twenty-one-year-old Herta Schiefelbein, who had come to look for a new position. The two women struck up an immediate rapport, and together they went to the Haberlandstrasse where Herta was shown the Einsteins' apartment. They agreed on terms of employment, and by lunchtime Herta was preparing lamb chops and green beans for the Einstein family, a dish that Albert pronounced the tastiest he had ever eaten.

This is how Herta came to join Einstein's household. Before long, she was a quasi-member of the family, which included Einstein's two stepdaughters and their husbands. She was cherished by them all, and she lived in their midst until 1933, when Hitler's accession to power compelled the Einsteins to forsake their two domiciles in Berlin.

Leaping fifty years into the future, we meet Herta again, in conversation with Friedrich Herneck, a historian of science, who had invited her to reminisce about her life in the Einstein ménage.[1] The year is 1978, and their five conversations take place at Herta's home in East Berlin, at the time the capital of the German Democratic Republic (DDR), or East Germany. In the five intervening decades, Herta Schiefelbein had married—she is now Herta Waldow—and had raised a son; Europe had been ravaged by the Nazis and by war, and the house in the Haberlandstrasse had been reduced to rubble in a bombing raid; but Herta's intelligence is as keen as ever, and her recollections of her life with the Einsteins are still vivid.

Herneck, her interlocutor, had a somewhat enigmatic past that reflects the political upheavals of twentieth-century Europe. After serving in the Wehrmacht in a noncombatant role, he found himself at war's end in a Russian prisoner-of-war camp—whether he defected or was captured is unclear. There he became an ardent Marxist, and, as an experienced stage actor and effective speaker, he was soon employed by his Soviet captors to re-educate his fellow prisoners. Following his release, Herneck lectured on dialectical materialism, the official Marxist philosophy, in various schools run by the DDR's Socialist Unity Party (SED), and finally at Humboldt University in Berlin where his course—three lectures a week—was mandatory for all students. East Germany was in those days an authoritarian state, and its citizens were indoctrinated in Marxist teachings from an early age. Although Herneck supported his country's political system, he remarked in his lectures on certain logical flaws in the official doctrine, and eventually this came to the attention of the authorities. Herneck was accused of "revisionism" (questioning the official dogma) and was abruptly relieved of his teaching duties. Thanks to the intervention of an influential friend, he was permitted to work on the history of science and was, in time, rewarded with a professorship in Berlin.[2] Herneck's scholarly interests were centered on the lives of notable nineteenth- and twentieth-century scientists, on their contributions and their philosophy, and he is the author of a number of carefully researched biographical books, several of them devoted to Einstein. But his work, published in the DDR, is little known in the West.

Chapter 4 of this book presents the five conversations in their entirety, in a translation that aims to preserve for anglophones the authentic flavor of Waldow's recollections. The copious notes of chapters 3 and 4 provide background information and brief biographies of the persons mentioned in the conversations.

Herneck's chief purpose was to draw out Waldow's recollections of her life with the Einsteins, and to convey thereby the atmosphere that pervaded their home. Waldow, for her part, perceived Einstein not as the celebrated scientist but as a kind and generous employer—the best she had encountered. She recalls how respectful and solicitous he always was toward her and her family, and how readily he was given to hearty laughter. We discover Einstein's favorite foods (heading the list are strawberries), what he liked to wear, his smoking habits, and who used to cut his hair; we also learn a good deal about Einstein's relationship with Elsa and with his women friends, and what provoked the occasional rows with Elsa. Waldow's story also reveals the enormous importance of music in Einstein's daily life, and we learn of his empathy with animals, specifically with a dog, a cat, and a parakeet. (Later on, in Princeton, he lived with a dog, numerous cats, and a parrot.) In short, Waldow gives us a pretty good idea of how Einstein spent his days in Berlin, from breakfast to night—particularly when her story is combined with Einstein's diary, in which he records his activities during a two-week period in 1931 (see chapter 3, the section titled "Two Weeks at Home").

On a larger scale, Waldow's recollections offer a glimpse of the social life of Berlin's academic and societal elite. We learn that the Einsteins often gave dinner parties for up to twenty-four guests, and we discover what dishes were served, who did the cooking, who peeled the asparagus, and even who washed the dishes. Unimportant as these minutiae may be, *in toto* they impart to us the flavor of Einstein's life in Berlin. Apart from formal dinners, the Einsteins also hosted informal gatherings and presentations to invited guests, for example, by Einstein, upon his return from a voyage. There was also a constant stream of visitors, both in the Berlin apartment and in nearby Caputh, the site of their summer home; some were occasional guests, such as the Bengali poet Rabindranath Tagore or Charlie Chaplin, and recurrent visitors from the conductor Erich Kleiber, colleagues such as Max Planck, and the cigar-smoking actress Hedwig Wangel to the American banker Henry Goldman—and dare one forget the "old Jewish man" who delivered fresh eggs to the Einsteins' apartment each week?

As a self-confessed aficionado of Einstein's life, Herneck could not resist injecting his own comments into his conversations with Waldow, usually to provide useful background information, and sometimes to present the results of his own research, such as what happened to Einstein's home and sailboat after 1933. From time to time, Herneck also raises ideological issues—some, evidently, for the benefit of the party officials who scrutinized his writings on behalf of the "head office for eternal truths," as Herneck liked to call it.[3] That this was a necessary ritual for authors in the DDR is borne out by the opening sentence of another book by Herneck: "Albert Einstein, one of the greatest Germans, after Karl Marx, occupies a unique position in the history of modern science."[4]

Herneck lived to see the Berlin Wall come down (1989). We do not know how he came to terms with the demise of the DDR, an event that must have had a wrenching effect on many academics and intellectuals who had invested heavily in Marxism and the DDR's official dogma, something not often considered. I am therefore very pleased that Professor Dieter B. Herrmann, an eminent historian of astronomy and a former student of Friedrich Herneck, agreed to contribute a short essay that addresses their dilemma. It appears in chapter 2.

Herta Waldow's recollections should be viewed against the backdrop of Weimar Germany in its last years; similarly, her reminiscences should be seen in light of the circumstances that brought Einstein to Berlin and made him a world celebrity. This background is provided in chapter 3, which also serves to introduce many of the personalities discussed in Waldow's conversations with Herneck in chapter 4. The first three conversations describe life in the Einsteins' city apartment, the fourth is devoted to their summer home in Caputh, and the fifth describes events in the aftermath of Hitler's accession to power.

CHAPTER 2

FRIEDRICH HERNECK, HISTORIAN OF SCIENCE IN DIFFICULT TIMES

by Dieter B. Herrmann

(Translated by Josef Eisinger)

I still recall very clearly my first meeting with Friedrich Herneck: It took place in 1956 in the Archenhold Observatory, when he and his wife were among the visitors on a guided tour that I conducted as a seventeen-year-old. The tour had attracted a goodly crowd because the planet Mars was unusually close to Earth and could be observed with considerable clarity. Afterward, we chatted together, and I discovered that he was the lecturer on "dialectical and historical materialism" at Humboldt University. When I told him that I would be starting my physics studies there in the coming term, he said as we parted, "Well, then we shall see each other again in the lecture hall."

And that is what happened. Herneck's lectures were fascinating for us students. They were not only very well prepared but were delivered with imaginative dramatic effects and impressive rhetorical flourishes. They were completely different from all the other lectures we heard. But the particular attraction of Herneck's discourses lay also in their content. Time after time, he raised questions about well-known propositions of dialectical materialism that had become dogma, by subjecting them to logical analysis. He raised doubts, for example, whether the discoveries of Soviet scientists could be explained by invoking dialectical materialism. "For how could one then explain the great successes of Planck and Einstein?" he would ask his breathlessly attentive audience.

An earlier publication of Herneck's had already earned him the displeasure of the official philosophers of the Party, that is, of the Socialist Unity Party or SED. After discovering in an archive an autobiography of the scientist-philosopher Ernst Mach, Herneck had published it together with some positive comments regarding Mach's controversial image. What made Mach's image "controversial" were negative comments that Lenin had made about Mach's epistemology in his 1908 book *Materialism and Empirio-Criticism*. Herneck had, furthermore, shown that some of Mach's ideas were part of the theoretical foundation of Einstein's relativity theory. Herneck's political higher-ups disapproved of such reflections because Lenin's writings about Mach were deemed to be the "ultimate wisdom." Herneck came under suspicion as much for the rejection of such dogmas as for his style of lecturing, as is borne out by the following excerpt from a 1962 evaluation of Herneck by the Institute for Marxism-Leninism: "His approach to teaching disorientated the students and led to demonstrative applause by negative elements among them."[1]

Herneck's case was by no means unique in those days. Beginning in 1957, a substantial number of philosophers at Humboldt University were designated "anti-Party and counter-revolutionary," and their interrogation had resulted in prison sentences of several years for some of them. These ideological struggles between the Party and intellectuals in the DDR must be seen against the political backdrop of the "de-Stalinization" then taking place in the Soviet Union in the wake of Nikita Khrushchev's famous "secret speech" at the Twentieth Communist Party Congress in 1956.

As far as we students were concerned, Herneck had simply vanished by fall of 1957. He had been dismissed, but because he was legally entitled to work the dismissal was revoked. Since his employment as a lecturer was now, however, out of the question, he was assigned the task of producing a three-volume history of Humboldt University, in celebration of the 150 years of its existence. In the course of this work, Herneck dealt extensively with the history of science, and eventually he became one of the most prominent historians of science in the DDR. His research on

Albert Einstein, Wilhelm Ostwald, Ernst Mach, and others also led to the publication of several successful popular science books that went through numerous editions, both in the DDR and beyond.

Herneck was born on February 16, 1909, in the ancient northern Bohemian town of Most (German name: Brüx), now in the Czech Republic. He had already been interested in science and philosophy as a high school student and, later, he added literature and dramaturgy to his interests and took courses in all of these subjects at the German University in Prague. He also worked as an actor in German theaters in what is now the Czech Republic, and wrote satirical pieces against the fascist Henlein party in pamphlets modeled on Karl Kraus's periodical *Die Fackel* (The Torch).

Following the German occupation of Czechoslovakia and the start of the Second World War, Herneck was called up for service in the Wehrmacht, in 1940. He joined the Nazi party (NSDAP) in the same year, presumably to be allowed to continue his dramaturgical studies while on leave from the Wehrmacht, and in 1941, he was awarded his doctoral title by the University of Erlangen.

By his own account, Herneck defected to the Soviet forces in 1945 and was confined in a prisoner-of-war camp. There he became acquainted with the writings of Marx, Engels, Lenin, and Stalin, and gave lectures to his fellow prisoners on politics in the spirit of dialectical and historical materialism. After his release, he was appointed lecturer on dialectical materialism at the College of Education in Potsdam, and, from 1954, at Humboldt University in Berlin. There his critical stance on political dogmas got him into serious trouble, as mentioned above. Thanks only to the staunch support he received from the chemist Robert Havemann, Herneck was allowed to work on the history of science and to defend his doctoral thesis on the chemist Wilhelm Ostwald. In time he was re-instated as a lecturer—this time, however, in the history of science. Havemann, on the other hand, fared much worse; he was declared an "enemy of the state" soon afterward and was placed under house arrest. Another five years passed before Herneck was appointed full professor, leaving him just seven years to teach at the University before his retirement. During that period he acted as expert examiner for my dissertation and also participated in supervising my doctoral thesis research.

The political demise of the DDR that took place in 1989–1990 (known as *die Wende*, the turnabout) affected Herneck profoundly, and not only in a positive way. His hopes that the DDR might become more rational were bitterly disappointed. Personal conviction had led him to invest a great deal of himself, and at considerable personal sacrifice, in the development and improvement of a state that had now, suddenly, vanished from the political scene. Einstein was evidently right when he wrote in his essay "Why Socialism?" "The attainment of socialism requires that some extremely difficult socio-political problems be solved: how is it possible, in view of the far-reaching centralization of political and economic power, to prevent the bureaucracy from becoming all-powerful and overweening? How can the rights of the individual be protected and thereby a democratic counterweight to the power of the bureaucracy be assured?"[2] A society that had evidently failed to find solutions to these problems had been the principal concern of Herneck's life for four decades. He had, nevertheless, never abandoned his faith that a better society than the present one would come to pass sometime in the future. Herneck died on September 18, 1993, in Berlin.[3]

CHAPTER 3

EINSTEIN'S ROAD TO BERLIN—AND BEYOND

by Josef Eisinger

It is hard today to grasp the magnitude of Einstein's renown in the 1920s and 1930s, when he was known and adulated throughout the world. The only parallel that comes to mind is with the Beatles in their heyday. Einstein's fame was thrust upon him in 1919, even though few of his admirers had any understanding of his scientific discoveries: their veneration had more to do with a universal feeling of relief that the horrific war had ended and a perception that relativity was something new, mysterious, and exciting.

In 1922, Einstein was invited to lecture on relativity theory in Japan, and this is the way Heinrich Solf, the German ambassador in Tokyo, described the public fascination with Einstein's visit in his report to the Foreign Office: "[Einstein's] journey through Japan resembled a triumphal procession. . . . The entire populace, from the highest dignitary to the rickshaw coolie, took part spontaneously, without any preparation or commotion. On Einstein's arrival in Tokyo, the crowd waiting at the station was so great that the police were powerless to control the life-threatening crowding. . . . Thousands of Japanese thronged to the lectures—at 3 yen per head—and [Einstein's] scholarly words were transmuted into yens that flowed into Mr. Yamamoto's [his host's] pockets."[1] Note that for three yen one could purchase ten meals at that time, and that each lecture lasted four hours, was given in German, and was translated, sentence by sentence.

What had Albert Einstein done to elicit such wild excitement? What were his origins? His passions? His achievements? And how did it happen that he found himself living in Berlin in 1914, as member of the Prussian Academy, when twenty years earlier he had forsaken his native Germany because he abhorred its militarism? These are among the questions that are addressed in this brief account of Einstein's personal and scientific lives. Readers who seek a more detailed narrative or additional historical sources are referred to the biographies listed in the select bibliography.

EARLY LIFE (1879–1895)

Einstein's ancestors, on both of his parents' sides, had long resided in the small towns of Swabia, the region in Southern Germany wedged between France and Switzerland. His father, Hermann, was born in 1847 in Buchau (now Bad Buchau), where a Jewish community had existed since the sixteenth century. Following the Napoleonic wars, when the regulations that severely restricted the domiciles and occupations of Jews gradually lapsed, the more enterprising among them migrated to the larger nearby towns, in search of greater opportunities. One of these was Hermann Einstein, who headed for the ancient imperial city of Ulm, thirty miles from Buchau. There he conducted a modest featherbed business, and in 1876 he married Pauline Koch, eleven years his junior, whose family was engaged in the international grain business. It was thanks to the far-flung members of the Koch family that, many years later, Einstein encountered cousins in Belgium, Spain, Argentina, and other places he visited in the course of his travels.

On March 14, 1879, the couple's son, Albert, was born in Ulm, and soon afterward the family moved to Munich, where Hermann and his younger brother Jakob, a graduate engineer, founded a company that manufactured motors and other electrical machinery. Two years after their move, Albert's sister was born and was named Maria, though she was always known by the diminutive Maja. She and Albert were to remain exceedingly close for the rest of their lives. According to reports, Albert did not enjoy playing with other children, preferring his own company, and was given to occasional temper tantrums. When he did find himself among other children, he was said to convey an aura of isolation—an aura that Einstein never shed entirely. He developed slowly and did not speak until he was about three, when he began to utter complete sentences—after having formulating them silently in his head.

Two events during Einstein's childhood were to affect his life profoundly. His father was responsible for the first when he presented his son—then about five—with a compass. According to Einstein's recollection, he was mesmerized by the compass because it demonstrated to him the existence of an invisible force field—in this case, the Earth's magnetic field that affected the compass needle in space. Later, he would greatly admire the British physicist James Clerk Maxwell for introducing into physics the electromagnetic field, as the concept of the "field" is at the heart of general relativity theory—and of the unified field theory, which Einstein searched for obsessively for much of his life.

Einstein's mother, Pauline, exerted the other powerful influence on him by introducing him to music. Herself an accomplished pianist, she arranged violin lessons for Albert when he was about six, but before long his teacher departed, after coming up against the boy's independence and distrust of authority. Einstein nevertheless became a competent, largely self-taught violinist, once he had been introduced to Mozart sonatas by his mother. After he became a famous scientist, he often surprised his hosts and reporters by arriving, as he usually did, carrying his violin case. While traveling, he was always on the lookout for chamber music partners, and when he was at home in Berlin he managed to play chamber music several times a week. Music was not just a diversion for Einstein, it stimulated his scientific thinking.

The first school Albert attended was a Catholic school in Munich where he was the only Jew in his class and where he experienced a mild form of anti-Semitism for the first time. His family, although staunchly nonreligious, adhered to the old Jewish custom of inviting a bible scholar to its weekly Sabbath meal—albeit with a nontraditional secular twist: their guest was a medical student, and he came to dinner on Thursdays. His name was Max Talmud and he, too, influenced young Albert profoundly when, on his visits, he brought the boy popular science books, as well as his first geometry book. Einstein read them avidly, and they helped him get over a brief interlude of intense religiosity that had been set off by the first Jewish religious instruction he received at his high school, the Luitpold Gymnasium. Talmud's science books convinced Albert that the stories in the Bible could not possibly be true and he became, according to his own account, a fanatical freethinker.

Einstein was, contrary to popular myths, a very good student, particularly in mathematics. All the same, he did not thrive at the Luitpold Gymnasium, where the emphasis was on Latin and Greek rather than on science. Einstein chafed, moreover, under the school's authoritarian teaching style, which stressed rote repetitions and discouraged students from asking questions.

At home, Einstein's family was going through uncertain times. The electrical machinery business ran into serious competition from larger companies located outside Munich, and in the end it failed. The family decided to move to Pavia, an ancient small town near Milan, there to make a fresh start. But Albert was to remain behind in Munich and to live with a distant relative, in order to finish high school and earn his Abitur (the university entrance certificate), the *sine qua non* for a member of the German bourgeoisie.

This arrangement was hardly to Einstein's liking. He missed his family, the militaristic culture that pervaded his school distressed him, and he dreaded the prospect of having to report for his compulsory military service in a little over a year. As he recalled many years later, he deemed it despicable then for any man to enjoy marching in formation. Einstein took matters into his own hands. He obtained a letter from Talmud's elder brother, a physician, stating that he was suffering from nervous exhaustion and needed to be with his family. The school was, apparently, only too pleased to let him go, for, as one of his teachers put it, Einstein's mere presence undermined the respect of the whole class for him. At the start of the Christmas vacation in 1894, Einstein took leave of his school and of his native land and appeared, unannounced, at the home of his astonished parents in Italy. The unrepentant high school dropout assuaged his parents' justifiable concerns by promising to study on his own and to prepare himself for the entrance examination of the Federal Polytechnic School in Zurich, known as Polytech (today's Swiss Technical University, or ETH).

Einstein stayed in his parents' apartment all spring and summer, reading up on physics while also taking a lively interest in the technical issues related to the family's electrical machinery business. As his sister recalls, he relished his new environment and the grace and natural ways of the local Italian population. He found time for long solitary hikes in the nearby mountains and for an excursion to Genoa, some fifty miles to the south, where he visited his uncle, Julius Koch. Einstein was still only sixteen—two years younger than most applicants—when he took the Polytech's entrance exam at the end of the summer. While he passed it easily in science and math, he was advised to improve his knowledge, especially in French and Chemistry, by attending for one year the cantonal school in Aarau, a short distance from Zurich.

This was a particularly fortunate piece of advice for Einstein. The teachers at the Aarau school were anything but authoritarian; they treated the students as individuals and were intent on helping them attain their goals; in short, the school was the opposite of the Luitpold Gymnasium. Einstein was also fortunate to board with the family of Jost Winteler, a teacher at the school, whose family made him feel so welcome that he shed his customary detachment and became virtually a member of the family. Jost Winteler's social philosophy was decidedly liberal, antimilitaristic, and antinationalistic, and he probably played a role in shaping Einstein's political views and reinforced his determination to renounce his German citizenship. Einstein, moreover, fell passionately in love for the first time, with Marie, said to be the prettiest of the Wintelers' daughters. Another daughter later became the wife of Michele Besso, Einstein's close and lifelong friend, while the Wintelers' son Paul eventually married Einstein's sister, Maja—making Einstein truly a member of the Winteler family.

STUDENT IN ZURICH (1896–1901)

After passing the final high school examination (Matura or Abitur) in Aarau with high marks, Einstein entered the Polytech in the fall of 1896. It was an exciting time in physics: X-rays and electrons had recently been discovered, and the concept of "quantized energy levels" would soon be introduced by Max Planck to explain the spectrum of radiating bodies (1900).[2] Thanks to Hermann von Helmholtz, Ludwig Boltzmann, James Clerk Maxwell, and Heinrich Hertz, the "new physics" was emerging. Einstein eagerly studied their work and came to the conclusion that the lectures of his professors at the Polytech were out of date—and skipped most of them. He also came to the mistaken belief that rudimentary mathematics was all that was needed for the formulation of physical laws and skipped the mathematics lectures of Hermann Minkowski as well. Ironically, it was Minkowski who, a few years later, gave relativity theory its elegant guise in terms of the space-time continuum, in which it is known to physicists today.

It is interesting that it was Heinrich Weber's physics laboratory that held the greatest attraction for Einstein, but he failed to endear himself to Weber—or to his other professors—by his intellectual independence and his sassiness (e.g., by addressing him as Herr Weber instead of Herr Professor). Einstein managed to obtain excellent grades, all the same, thanks to the meticulous lecture notes he borrowed from his fellow student and friend, Marcel Grossmann. In the next dozen years, two more occasions would arise on which Grossmann would come to Einstein's rescue in very consequential ways!

Einstein's personal student life was bohemian in style, by both necessity and inclination. His mother's family, the Kochs, provided him with a modest monthly stipend of SF100, and he lived in a plain rented room. For recreation, he smoked his pipe in the company of fellow students, as they discussed philosophical questions at the Café Metropole. Apart from physics, it was music that afforded him the greatest personal inspiration, as it would for the remainder of his life, and he missed no opportunity to find a pianist for Mozart sonatas or to enlist string quartet partners.

During Einstein's last two years at the Polytech, while writing a thesis and under pressure to prepare for the final exams, he was also subject to a powerful infatuation with a fellow student, Mileva Marić. Mileva came from Vojvodina, a region of Serbia that had been part of the Habsburg monarchy since the early eighteenth century. Her father, an official in the Hungarian civil service, had provided Mileva with an excellent education, sometimes in schools where she was the only girl in the class. Mileva had come to Switzerland to finish high school, and after she passed her Matura in 1896 she registered as a medical student at Zurich University, one of the few universities that admitted women at that time. But she soon transferred to the physics section of the Polytech, where she hoped to earn the diploma qualifying her to teach mathematics and physics in high schools. It was there that her path crossed Einstein's.

Mileva Marić and Einstein became friends. She accompanied him on the piano—as Marie Winteler had done—but there the similarity between the two women ended. Contemporaries described Mileva as being small, moody, and silent, and walking with a slight limp. Einstein's friends were surprised that this good-looking, lively young man was so strongly attracted to her, but he evidently saw in her a soul mate with whom he could share his burgeoning ideas about physics. They studied together, shared their frugal student meals, and, as their ample correspondence shows, they declared their love for each other. The only cloud over their blossoming relationship—and it was a dark one—was the vehement opposition of Einstein's parents, once he told them of his intention to marry Mileva.

Einstein and Mileva's final year at the Polytech was dominated by the looming final examinations in the summer of 1900. Out of the five candidates, four, including Einstein, passed and received their teaching diplomas, but Mileva did not. However the assistantship that would have placed Einstein's foot on the first rung of the academic ladder, and that he had confidently counted on, did not materialize. At the very time when his ideas in physics were maturing in his mind, he was faced with formidable new challenges in his personal life.

Having failed the examinations, Mileva decided to study at her parents' home in Novi Sad before taking them again in the following year. Einstein, meanwhile, remained in Zurich to work on a doctoral thesis. Occasionally, he traveled to Pavia and stayed with his parents—enduring violent confrontations with his mother on account of his continuing liaison with Mileva. During the winter, Mileva joined Einstein on a passionate joint holiday, which they spent at Lake Como and its exquisite surroundings.

Throughout that year, Einstein searched tirelessly, but fruitlessly, for an assistantship at a university, and he became convinced that his singular lack of success must be ascribed to his antagonistic relations with Professor Weber. To earn money, he accepted temporary teaching jobs, while always staying in touch with the latest developments in physics. When Mileva discovered that she was pregnant, Einstein was very supportive and loving in his letters to her, even as he kept her apprised of the progress he was making in his own work. Mileva, meanwhile, was under great stress and failed her diploma examinations at the Polytech a second time. She returned to Novi Sad, where their daughter was born in January 1902. In their letters, Mileva and Einstein referred to the baby as "Lieserl," short for Elisabeth, yet all that is known of her is that some time later she was given up for adoption.

Thanks to the voluminous correspondence between Einstein and Mileva, their states of mind during this dramatic and uncertain period of their lives are known in remarkable detail, and interested readers are referred to more exhaustive biographies.[3]

PATENT CLERK IN BERN (1902–1908)

Einstein was working as a tutor in Winterthur, a small city not far from Zurich, when he received the good news that the job at the Federal Patent Office in Bern, a job that his friend Marcel Grossmann had helped to arrange, had come through for him. Einstein quit his tutoring job precipitously and moved to Bern, although he had to wait until June 1902 before his official appointment as Technical Expert Class 3 was confirmed. His yearly salary was now a respectable SF3,000.

To earn money in the interim, Einstein placed an advertisement in the newspaper that offered private lessons in mathematics and physics, with a trial lesson, gratis. One person who responded to the ad was Maurice

Solovine, a student at Bern University with whom Einstein developed an immediate rapport and who became a life-long friend. They were soon joined by Conrad Habicht, also a university student, and these three men formed, mock-seriously, the Olympia Academy, with Einstein as its president. Einstein was in his element. He now had a platform for honing his ideas on physics and was among admiring friends. The Academy, which soon expanded to include three additional members, had a remarkably erudite reading list that included the works by Immanuel Kant and Ernst Mach, besides David Hume's *A Treatise of Human Nature*, Sophocles's *Antigone*, Cervantes's *Don Quixote*, Spinoza's *Ethics*, and Poincaré's *Science and Hypothesis*. In his letters to Mileva, Einstein described the stimulating discussions of the Academy and also the carefree outings of these high-spirited young men. Mileva, meanwhile, was recovering from a difficult child birth in Novi Sad and was in sore need of the fervent reassurances that Einstein sent her.

In the autumn of 1902, Hermann Einstein fell seriously ill with heart disease, and Einstein rushed to his bedside. He was devastated when his father died soon after. Because of his father's repeated business failures there were no savings, and his mother, Pauline, spent the rest of her life as a housekeeper in the homes of others—for the longest time in the home of her sister, Fanny, and her brother-in-law, Rudolf Einstein, the parents of Elsa, Einstein's second wife. Hermann, on his death bed, gave Einstein his permission to marry Mileva, but Pauline never relented in her hostility toward her.

Einstein was now a salaried Swiss citizen, and early in 1903 he and Mileva were married in the registrar's office in Bern. The ceremony was attended by members of the Olympia Academy, with no Einstein or Marić family members present, and it was followed by a communal meal in a local restaurant. Except for the shadow cast by the Lieserl episode, the newlyweds lived harmoniously together during the next several years, with Mileva often attending the meetings of the Olympia Academy, but without participating in them.

Einstein had been warned that his job at the patent office would bore him, but this did not turn out to be the case: his duties consisted of examining the validity of claims made in patent applications, and he found the work varied and challenging. His fascination with inventions and patents continued long after he left the patent office, and he was, eventually, the owner of several patents. Most important, his position allowed him to keep up with the latest developments in physics, and it was during his early years at the patent office that his first scientific papers (on thermodynamics and statistical mechanics) appeared in the prestigious journal *Annalen der Physik*.

In May 1904 the couple's first son, Hans Albert, was born, and the happy event prompted Mileva's elated father, Miloš Marić, to travel to Bern to see his grandson. The following year, 1905, has been dubbed Einstein's *annus mirabilis* because in a period of just a few months he published four articles on disparate subjects that can be seen as lifting the curtain on "modern physics":

(1) The special theory of relativity, which replaced Newton's concept of absolute time and space and obviated the existence of the ether. It also established the equivalence of mass and energy ($E = mc^2$) and was therefore a harbinger of nuclear physics, still far in the future. It would take Einstein another ten years to extend the theory to include gravitation (the general theory of relativity).

(2) A molecular theory of liquids that allowed Einstein to derive the size of molecules from the measured diffusivity and viscosity of a liquid—this at a time when some scientists still questioned the existence of molecules!

(3) A molecular theory of heat that explains the random motion of small particles suspended in a liquid, a phenomenon that had long been known as Brownian motion.

(4) A corpuscular theory of light that explains the previously puzzling properties of the photoelectric effect—the ejection of electrons from a metal surface irradiated by light, and its dependence on the wavelength of the light. This is the work for which Einstein was awarded the Nobel Prize in 1922, largely because his relativity theory was still considered to be too controversial by the members of the Swedish Academy.

What these four papers have in common is that they clarify phenomena that classical physics was unable to explain, and Einstein had accomplished this by viewing each in an entirely new way. So new that even eminent physicists, including Planck, remained skeptical for a long time about Einstein's proposition that light has the characteristics of both a wave and a stream of quanta of radiation (or photons, as they are now called). But before long, as experiments validated Einstein's theories and theorists recognized their significance, his reputation in the physics community grew, although he remained largely unknown among the general public.

ZURICH AND PRAGUE (1909–1913)

Following his *annus mirabilis*, the scientific output of the patent examiner continued unabated. Among that output was a quantum theory of solids that explained the "anomalous" behavior of the specific heat of solids at very low temperatures. The experimental data that Einstein's theory was able to explain had been obtained by Walther Nernst in Berlin, who now became one of Einstein's earliest champions.[4] In 1908, the University of Bern appointed Einstein *Privatdozent*, an unpaid position that required him to give seven lectures a week—still, it was his first academic appointment. Einstein was now obliged to juggle his lecture schedule with his hours at the patent office, with the result that his lectures often took place at 7:00 a.m. and were generally attended only by the members of the Olympia Academy and Mileva.

A year later (1909), the University of Zurich offered Einstein an associate professorship, that led him to resign his position at the patent office and to move back to the city that he and Mileva had known and loved as students. Although his salary was the same as in Bern, their living costs were now considerably higher, and when the couple's second son, Eduard, was born in the following year, they were obliged to take in a boarder to make ends meet.

In Einstein's professional life, things were looking up, however. In recognition of his various remarkable contributions to physics, he was invited to address a large and important scientific meeting in Salzburg, which gave him his first opportunity to meet some of the giants of theoretical physics—among them Max Planck, Arnold Sommerfeld, and Max Born.[5]

In Zurich, Einstein was burdened with a considerable teaching load, which, along with his other academic duties, left him less time for research than he had enjoyed in Bern. As a result, he was under substantial stress, and Mileva's letters reveal that she felt increasingly neglected.

Einstein's career now took flight. In 1911 he received an even more prestigious appointment—the chair for theoretical physics at the German University of Prague. His salary was now large enough for the family to adopt a proper bourgeois lifestyle in a large apartment, and even to employ a maid. Nevertheless, neither Einstein nor Mileva felt at home in that city. In letters to his Swiss friends Einstein complained about the onerous bureaucracy, the paper work he encountered in Prague, and his students' lack of interest in his subject. He was also distressed by the mutual hostility between the Czech- and German-speaking citizens of Prague and both he and Mileva missed their beloved Zurich. Einstein's friends in Switzerland went to work, and in January 1912 their efforts bore fruit: Einstein was appointed professor of theoretical physics at the Polytech—now known as the Federal Institute of Technology, or ETH. The prospect of returning to Zurich delighted Mileva and the two boys as much as Einstein, and it is safe to say that the irony of this exalted appointment, ten years after his alma mater had denied him a lowly assistantship, was not lost on Einstein.

The comparative academic isolation Einstein endured in Prague allowed him to revisit a problem that had occupied him for several years. He was well aware of the "special" character of the relativity theory of 1905, in that it applied only to systems whose relative motion was uniform and linear, while excluding accelerated (or rotational) motion. Einstein's insight that for a person falling freely in the gravitational field, gravity would cease to exist—what he later called the happiest thought of his life—led him to apply to accelerating systems the same "principle of equivalence" he had employed in special relativity; to wit, that in each system, all physical laws and the velocity of light are the same.

But in implementing these two postulates, Einstein was confronted with mathematical difficulties he was unable to overcome. His move back to Zurich therefore took place at a most opportune time, for his old friend Marcel Grossman, who was now professor of mathematics at the ETH, came to his rescue for the third time. He recognized that implementing the principle of equivalence led to a space-time manifold whose geometry was not Euclidean but of a type that had long been thoroughly investigated by mathematicians—beginning with the great Friedrich Gauss at the beginning of the nineteenth century. Einstein gratefully followed Grossmann's lead and delved into a study of non-Euclidean geometry. It would, nevertheless, take him another two years of hard work before he completed his greatest achievement, the general theory of relativity, which demonstrated that space, time, and mass were inseparable from each another.

As Einstein's renown among physicists grew, so did Mileva's isolation, and her hopes that their return to Zurich would rekindle the happiness they had known in their student days were sorely disappointed. Einstein continued to be an attentive, loving father, but the preparation of lectures and his other obligations, while grappling with his mathematical difficulties, sapped his energy and strained their marriage.

But an even greater threat to their marriage loomed for Mileva when two emissaries, Max Planck and Walther Nernst, arrived from Berlin to make Einstein an offer that was designed to be impossible to refuse. The package they offered consisted of membership in the prestigious Prussian Academy of Sciences, a professorship at the University of Berlin, and the directorship of the planned Kaiser Wilhelm Institute for Physics, combined with a handsome salary and best of all with no teaching obligations whatever. Berlin was then Europe's premier center for scientific research, where Einstein would be surrounded by colleagues who were among the brightest stars of contemporary science. Einstein deliberated for only one day before accepting the offer.

Apart from the undoubted academic attractions Berlin offered, the city offered another that was not known to the two emissaries. On his first visit to Berlin in 1912, Einstein had re-established contact with his cousin Elsa, whom he had known since childhood and who, newly divorced, lived with her two daughters in Berlin. During his stay in that bustling metropolis,

Einstein had enjoyed her cheerful company on several joint outings, and after his return to Prague he and Elsa entered into an affectionate and wistful correspondence.

In September 1913, Einstein, Mileva, and their two boys traveled to Novi Sad in Serbia to visit her parents in their summer home, and while there the strains on their marriage were glossed over. Einstein then journeyed to Vienna for a lecture, and to Berlin to visit his mother and his other relatives, Elsa among them. Mileva was understandably apprehensive of the impending move to Berlin that would bring her into close proximity to Einstein's relatives, and also about exchanging her beloved Zurich for a place known for harboring prejudices against Slavs. Einstein's letters to Elsa, meanwhile, grew more intimate, and in them he complained of Mileva's moodiness. Nonetheless, his involvement with Elsa was by no means the main motive for his move to Berlin. The motive was certainly the incomparable working environment he enjoyed there, but Elsa's companionship probably made the prospect of life in the Prussian capital more bearable.

BERLIN (1914–1932)

The War Years, Divorce, and General Relativity

At Christmas time in 1913, Mileva traveled to Berlin to look for suitable housing for the family, while carefully avoiding all contacts with Einstein's family, and especially with Pauline. While in Berlin, Mileva stayed at the home of the eminent physical chemist Fritz Haber and his wife, Clara.[6] The Habers had taken a liking to Mileva and helped her find an apartment in Dahlem, the suburb where the Kaiser Wilhelm Institute for Physics was planned to come into existence. They found a suitable one at Ehrenbergstrasse 33, and in the middle of April, after the renovations were completed, the Einstein family moved into their new home.

Within two months, however, Mileva's apprehensions were realized. The tensions related to Einstein's liaison with Elsa and Pauline's undiminished hostility toward Mileva came to a head, and the marriage, which had been fraying for some time, was now in tatters. At the end of the school year Mileva took the two boys back to Zurich, where she rented rooms in a boarding house. Haber who had accompanied Einstein to the railway station to see them off, related that Einstein was badly shaken during the farewell, and that he broke into tears on the way home.

During the following four years, Einstein and Mileva carried on a voluminous correspondence, marked in the beginning by mutual distrust. The efforts of Fritz and Clara Haber and others to reconcile the couple came to naught. Einstein accused Mileva of turning Hans Albert against him, while Mileva complained of being financially strapped. By the end of the 1914, Einstein had agreed to send Mileva a yearly stipend of RM 5,600, claiming that this left him with only a bare minimum for himself, sufficient for a room in which to live and work. This stipend allowed Mileva to rent a flat in the vicinity of the ETH, and she supplemented it by giving private lessons in physics and in piano. She gradually regained control over her long-distance marriage because Einstein needed her cooperation for staying in close touch with his sons—for visits with them, for walking tours in the Swiss mountains, and for sailing holidays in the Baltic Sea.

In spite of Einstein's intense involvement with family matters and physics, he could not have failed to be aware of the momentous world events unfolding at the same time. Following the assassination of Archduke Ferdinand in Sarajevo in July 1914, Austria declared war on Serbia and, like dominoes in a row, Russia, Germany, France, and Britain quickly followed suit and went to war. Although Einstein was appalled by the enthusiasm with which the conflict was greeted by even his closest colleagues, the war had little impact on his work habits. Nor did the war affect his close relationships with Haber and Nernst, who were developing chemical weaponry and played active roles in deploying the first poison gas attack on the Western front. The war years were, indeed, a very productive period for Einstein: Between 1914 and 1918 he published no fewer than thirty scientific papers on a variety of subjects.

Einstein also found time to give science talks to nonspecialist audiences, something that was always close to his heart. Indeed, his first public presentation of the general relativity theory was to a lay audience and took place in June 1915 in the auditorium of the Treptow (now, Archenhold) Observatory. Einstein had been invited by the astronomer and founder of the observatory, Friedrich Archenhold—who would become a frequent visitor at Einstein's Haberlandstrasse apartment in the years to come.[7]

Following the break-up of his marriage, Einstein moved to a more centrally located apartment on the Wittelsbacherstrasse, where he lived alone in a sparsely furnished apartment for about three years. In the summer of 1915 he and Elsa spent a vacation together on the wild, remote island of Rügen in the Baltic Sea, and his letters to Swiss friends reveal that he found his living arrangement—conveniently close to Elsa's apartment—to his liking.

The tense exchanges between Einstein and Mileva following their separation took place while he was working strenuously on completing the general theory of relativity—evidence of Einstein's extraordinary powers of concentration. A breakthrough came toward the end of 1915, when he was able to demonstrate that the theory accounted for a phenomenon that had puzzled astronomers since it was first reported in 1859: The perihelion of Mercury (the planet's position when closest to the sun) had been observed to shift slightly with each orbit of the planet, and the magnitude of this shift—43 seconds of arc per century—now turned out to agree with that calculated by relativity theory. Einstein was elated by this agreement, but the world at large had to wait four more years, before a different astronomical observation also confirmed his theory—and made him a world celebrity.

In 1916, Einstein proposed a divorce settlement to Mileva, possibly upon the urgings of Elsa and her family. His approach precipitated a violent confrontation, and Mileva suffered a mental and physical breakdown. In the following year, Einstein also fell seriously ill with digestive problems. He lost a great deal of weight, probably due to the severe food shortages in Berlin and his poor wartime diet. He recovered thanks to Elsa's devoted nursing care, and thanks to her access to the scarce foodstuffs she was sent by her relatives in Swabia; soon afterward (1917) Einstein gave up his bachelor quarters and moved into the spacious Haberlandstrasse apartment where Elsa lived with her two daughters from her first marriage.

Following that move, Einstein again approached Mileva to ask for a divorce settlement. Mileva, who had barely recovered from her illness, was grappling with another family problem at that time. Her sister, Zorka, who had come to Zurich to take care of her, became mentally ill and was admitted to the Burghölzli psychiatric clinic—the same clinic in which Einstein's younger son would later reside. Following lengthy negotiations, conducted on Einstein's behalf by his tireless friend Fritz Haber, a settlement was agreed upon: Mileva was to receive an alimony amounting to more than half of Einstein's salary, as well as all of the Nobel Prize money—once the prize was awarded to him. Both Einstein and Mileva were evidently confident that the prize would eventually come to him, and a few years later it did. The final divorce decree was issued in February 1919, and in June Einstein married Elsa.

Fame, Politics, and the Gyrocompass

In the same year, an event took place that profoundly affected Einstein's future life. Sir Arthur Eddington, the director of Cambridge Observatory and Britain's principal exponent of the general relativity theory, obtained funds from the British government for two scientific expeditions, with the mission of determining if the gravitational field of the sun deflects light by the amount predicted by general relativity theory.[8] Since the requisite observations were feasible only during a total eclipse of the sun, the expeditions, equipped with telescopes and cameras, traveled to two locations where the 1919 eclipse would be observable: one in Brazil and the other on Principe, an island off the west coast of Africa. There the astronomers recorded the positions of certain stars on photographic plates in the absence, and again in presence of the eclipsed sun, and upon their return to England the plates were carefully analyzed. Then, in November 1919, in the rooms of the venerable Royal Society in London—the publishers of Newton's *Principia* in 1687—Eddington proclaimed with great pomp that the experiment had verified Einstein's theory.

The response of the public was immediate and overwhelming. Before Eddington made his announcement, only physicists were familiar with

Einstein's name; afterward, he was known and admired in every corner of the world. Newspaper headlines everywhere proclaimed the demise of Newton and the triumph of Einstein, and it mattered little that few of his devotees understood the import of his discovery. It did not hurt that Einstein was a choice photographer's model, that he had a charismatic personality, and that he was not awed by authority. The new media of radio, newsreel, and film disseminated his image far and wide, and his views on matters that had nothing to do with science were eagerly sought by reporters. It seemed that his mysterious theory helped erase the memory of the four years of senseless slaughter that had just ended.

Einstein was astonished by his fame. To his friend Harry Kessler, the so-called Red Count, he confided that he understood why Copernicus had produced a stir when he deposed the Earth from its central role in creation, but *his* theory did not at all affect the way people perceived their place in the universe; it made him feel like a con artist who had failed to give people what they expected from him. Kessler, a perceptive observer, also commented on Einstein's ever-present slightly ironic facial expression, saying that he reminded him of someone smiling at human conceit, not just superficially but down to its very roots.[9]

As time passed and the public's fascination with Einstein failed to subside, he learnt to use his fame to promote the humanitarian causes that were dear to his heart. Though he often proclaimed himself a loner, he could not escape his irrepressible sense of social justice. Countless organizations sought—and enjoyed—his support, from Zionists to pacifists and groups promoting friendship with Soviet Russia or the education of workers. While he deplored being constantly pursued by reporters and photographers, another side of him appreciated the benefits of being a celebrity; among them, financial independence, access to the rich and famous, and invitations to lecture tours that allowed him to travel all over the world.

The end of the First World War also ended the reign of Kaiser Wilhelm II. In the wake of the armistice, sailors mutinied in Kiel, and in Berlin soldiers and workers hoisted a red flag over the Imperial Palace. Out of this turmoil, the Weimar Republic was born, its survival under constant threat from both the extreme Right and Left. Those on the Right, and particularly the Nazis, championed the so-called *Dolchstosslegende*, the myth that a "stab in the back" by perfidious politicians and Jews had led to Germany's defeat in the war. For them, Einstein's fame and his beliefs made him an obvious target.

In August 1920, Einstein encountered the first serious manifestation of the virulent anti-Semitism that permeated nationalist parties. Paul Weyland, a member of such a party—and a confidence trickster—obtained financial backing to organize a large rally in Berlin's Philharmonic Hall at which Einstein's theory was attacked as a hoax, also on account of its "Jewishness," and because it was plagiarized, to boot. This event and its sequels in the press shook up Einstein, but worse was to come. Two years later, the Republic's foreign minister, Walther Rathenau, long vilified as a Jew, was assassinated by two young army officers, and just one week after that, the Jewish journalist Maximilian Harden, a fierce critic of Wilhelm II, was also attacked and seriously injured. Einstein had been acquainted with Rathenau and, like him, had often been warned by the police of being in serious danger. He took heed, withdrew from his most controversial activities, and even hatched a plan to abandon academia entirely and to leave Berlin.

Though Einstein would soon discard this drastic scheme, the episode sheds light on his frame of mind at the time, as well as on the breadth and depth of his interests. After abandoning his career as a patent examiner, he had always retained, maybe a little wistfully, a lively interest in inventions and technology. In 1914, while serving as a court-appointed expert witness in a patent dispute, Einstein met Hermann Anschütz-Kämpfe, the founder and owner of a company that manufactured gyrocompasses in Kiel. The two men became close friends and began to collaborate on improving the design of gyrocompasses, work for which Einstein would be awarded several patents. One needs to recall that at that time, long before the advent of global positioning satellite (GPS) technology, the gyrocompass was, by far, the most advanced navigational instrument available and was widely used in both ships and planes.

Now, in the wake of Rathenau's murder—which had affected him deeply—Einstein proposed to his friend Anschütz that he give up his academic positions in Berlin and work as an engineer in Anschütz's firm in Kiel.[10] Anschütz was flabbergasted at this suggestion and made a counter proposal: Einstein should retain his positions in Berlin, and he, Anschütz, would put at his disposal, as a refuge, a private residence in Kiel, which was on the water and equipped with a sailboat. There Einstein could stay whenever he needed to escape the overcharged political atmosphere in Berlin. Einstein was delighted with this arrangement and made frequent use of it in the years that followed. We know from his letters, for instance, that in the summer of 1923 he enjoyed a lengthy stay at his "lavish Kiel haven," that together with his son he had sailed out to the open ocean where the swell was considerable—and that it was "wonderful."

Zionist and Voyager

With such a luxurious sanctuary to fall back on, Einstein settled again into his comfortable bourgeois existence in Berlin, with Elsa ably taking care of his day-to-day needs. As their relationship grew less passionate and more comradely with the passing years, she even came to accept, albeit reluctantly, his friendships with other women. Elsa was rightly confident that he would never leave her and found contentment in being married to this famous and exceptional man.

Rising anti-Semitism in Germany raised Einstein's awareness of being a "member of the tribe," as he liked to put it. Although he never had any desire to live in the Jewish homeland, he recognized that Zionism offered a practical solution for the plight of Europe's Jews; but even as he gave the movement his wholehearted support, he despised all nationalism and never failed to admonish the Zionists to include the Palestinian Arabs in their plans for a homeland. In the spring of 1921, Chaim Weizmann, the newly elected president of the World Zionist Organization, asked Einstein to join him on a fundraising trip in America in support of the planned University of Jerusalem (now, Hebrew University). Einstein agreed and was accorded tumultuous receptions wherever he went.

A year later, a Japanese magazine editor invited Einstein to an extended lecture tour in Japan. Einstein recognized the invitation as a wonderful opportunity to escape the ferocious politics at home and to explore a fascinating, little-known land and culture—and he accepted.

A journey to the Far East was no small undertaking in that era (1922). After embarking in Marseille, Einstein and Elsa were confined for a month on a modest mail steamship (8,500 tons), much of the time in tropical and, sometimes, stormy weather—naturally, without air conditioning. Along the way, they visited Ceylon (Sri Lanka), Singapore, Hong Kong, and Shanghai, before reaching Japan where an enthusiastic reception awaited them. After spending two months in Tokyo and several other cities, they began their return voyage on a similar ship that again made a number of stops along the way.

Einstein had been invited to spend two weeks touring in Palestine, then still a British mandate, and when his ship reached Port Said, the northern terminus of the Suez Canal, he and Elsa disembarked and travelled to Jerusalem by train. His Zionist hosts made strenuous, but unsuccessful, attempts to win him for the nascent University of Jerusalem, but, as he confided to his diary, "the heart says yes, but reason says no." The final destination on Einstein and Elsa's long odyssey was Spain, where they were again feted elaborately by scientists, Spanish royalty, and the public. They returned to Berlin after an absence of almost six months.

Einstein usually kept a diary when on a major voyage and used it to record his experiences and his impressions of people he encountered. These travel diaries [*Reisetagebücher*]—the only diaries he kept—served as useful memory aids in the presentations Einstein gave to invited friends, after returning from a journey.[11]

Einstein treasured the tranquility and solitude afforded by sea voyages, and two years after returning from the Far East he went to sea again. He had repeatedly been invited to lecture in South America, and in the spring of 1925 he agreed to a lecture tour that took him to Argentina, Brazil, and Uruguay. Margot Einstein, the younger of his stepdaughters, was to accompany him on this trip, but she fell ill shortly before the departure date, and in the end Einstein traveled alone. Once aboard ship, he managed to recruit two fellow passengers and the ship orchestra's first violinist to join him each morning in playing string quartets in his steamy cabin. As he commented in his diary, "one sweats a lot, but it is delightful."

He was, however, already dreading what awaited him upon his arrival, and his fears were not unjustified. A crowded schedule of lectures and receptions had been planned for him, and while he was irritated by the shallowness and loquaciousness of some of his hosts, he felt most at ease in Uruguay, a country whose political system reminded him of Switzerland. After his final goodbye in Brazil, he confessed to his diary: "free at last, but more dead than alive." Even after recuperating somewhat during his homeward voyage, he returned to Berlin in a state of exhaustion, and his doctors urged him to abstain from similar ventures for a few years.

But it did not take long before Einstein plunged back into the spirited political and social life of Berlin. He and Elsa were often invited to dinner parties at which they mingled with nobility, journalists, and members of the city's financial, diplomatic, intellectual, and artistic élite.

In June 1927, Herta Schiefelbein joined the Einstein ménage as housekeeper, and she soon became an integral part of it. At about the same time, Hans Albert, now twenty-one, announced to his parents that he intended to marry Frieda Knecht, a woman nine years his senior. He met with strenuous objections from both Einstein and Mileva—in almost a reenactment of the opposition encountered by Einstein when he told *his* parents of his plan to marry Mileva twenty-five years earlier. Even the intercession of Einstein's friend, Anschütz, failed to change Hans Albert's mind, and the two lovers entered into matrimony. It turned out to be a long and, by all reports, successful marriage, and both Einstein and Mileva were soon reconciled to it.

Hans Albert's younger brother, always of a more artistic and introspective nature, fared much worse. Eduard, "Tetel" as Einstein called him, studied medicine at Zurich University, but he often slipped into a depressive state and had attempted suicide. His condition worsened gradually, in spite of the combined best efforts of his parents. He was eventually diagnosed with schizophrenia and spent the remainder of his life in the Burghölzli psychiatric institution in Zurich. Einstein was profoundly distressed by Eduard's fate and felt certain that Tetel had inherited the illness from Mileva.

Einstein at Fifty: Caputh and Pasadena

On March 14, 1929, Einstein turned fifty, and the city council of Berlin wished to honor its most illustrious citizen by presenting him with a plot of land near water where he could build a summer home and indulge in his favorite pastime of sailing. The idea for this gift had been suggested to the council by Einstein's doctor and friend, János Plesch, and at first Einstein agreed enthusiastically to it.[12] Unfortunately, the actual transfer of the gift turned into a comedy of errors, since the city administration had neglected to ensure that the offered site was theirs to give. In the end, Einstein informed the city council that things had gone on far too long and that he was declining the offer. He may also have realized that the city's gesture would inevitably result in public controversy. In any case, he acquired a plot on his own. It was located in the village of Caputh near Potsdam, on rising ground and within walking distance of the lakeshore. The site looked out over the Templin and Schwielow Lakes, part of the Havel chain of lakes, and above it stretched a large state forest in which Einstein took solitary walks and gathered mushrooms.

Once the idea of a summer home had been planted in his head, Einstein wasted no time getting his dream house built. He hired a young architect, Konrad Wachsmann, an employee of a large timber construction firm, and within a few days he designed a modest, functional summer home for the Einsteins. The prefabricated timber building had small bedrooms, two decks, and a large living room with French doors that presented a vista of the lovely lake scenery below. For the summer of 1929, the Einsteins rented an old house in Caputh that was within walking distance of the building site and this enabled Elsa to check on the progress of the construction. The foundation of the house was completed in July, and by October the family, along with the indispensable Fräulein Herta, moved into their new home.

In the same summer, a handsome twenty-three-foot cabin sailboat with a mahogany deck and gleaming brass fittings was delivered to Einstein in Caputh by its builder, in person: this was the *Tümmler* (porpoise), the magnificent fiftieth birthday present from a group of wealthy friends, among them the American banker Henry Goldman.[13] During the four summers Einstein owned the *Tümmler*, he spent untold hours sailing on it, usually alone, and sometimes with select friends, such as Erwin Schrödinger or Toni Mendel.[14] During one summer, his weekly sailing companion was his Austrian friend Margarete Lebach—one visitor Elsa took pains to avoid by leaving for the city early in the morning and returning late in the evening.

The *Tümmler* was surely his most imposing birthday present, but Einstein received many others, from the humblest and the grandest of his admirers, and he was particularly touched by the small presents sent by children. To avoid the onslaught of well-wishers, he spent the day in seclusion at his hideaway on Plesch's estate in Gatow. In the afternoon Elsa and Margot joined him there for a private birthday celebration.

Einstein treasured his Caputh home for its tranquility—calling it a paradise in a letter to his sister—yet it was also a Mecca for a stream of visitors. The most exotic among them was the Bengali poet, musician, and philosopher Rabindranath Tagore, who enjoyed tremendous popularity in Germany during the 1920s. Many people perceived Tagore and Einstein as the spiritual representatives of East and West, and their dialogues were published and widely read.[15]

Einstein traveled extensively in Europe, but he did not embark on another sea voyage until December 1930, when he, his secretary Helen Dukas, and Elsa boarded the SS *Belgenland* in Antwerp. They were on their way to America to spend the winter session at the California Institute of Technology (Caltech) in Pasadena, California. It is not surprising that Einstein chose to travel there by way of Cuba and the Panama Canal, the route that afforded him the greatest number of days at sea. The *Belgenland* had a four-day stopover in New York, where Einstein's arrival caused a hullabaloo that exceeded anything he had imagined in his "most fantastic expectations." During his brief stay in the city, he survived a gigantic Zionist rally, a reception at City Hall, and a speech comparing him to Copernicus; he also attended a performance of *Carmen* at the Metropolitan Opera, where he was warmly applauded by the audience when he and Elsa were spotted in their box, and he met with Fritz Kreisler, Arturo Toscanini, and John D. Rockefeller, among many other dignitaries. Back on board, as his ship pulled away from the dock and headed for sunny Havana, Einstein confided his "great sense of liberation" to his diary. After short stays in Cuba and Panama City, Einstein and Elsa landed in San Diego and were driven to Pasadena.

After the stiff formality at the Prussian Academy, Einstein was exhilarated by the congenial and relaxed atmosphere of the academic gatherings at Caltech. Most notably, he visited Edwin Hubble at the Mount Wilson Observatory, who showed him the photographic plates on which the spectra of distant galaxies were recorded—the spectroscopic evidence that the universe is expanding. Hubble's momentous discovery prompted Einstein to renounce the cosmological constant that he had introduced in his field equation to account for the universe being static—as was then believed. At a formal reception in the Athenaeum, Einstein thanked Albert Michelson for "abolishing" the ether with his famous experiment at a time when he, Einstein, was just three feet tall.[16] While at Caltech, Einstein also visited the novelist and social critic Upton Sinclair, witnessed the Rose Bowl parade, and found a willing chamber music partner in the once-celebrated violinist Lili Petschnikoff. Since Hollywood is not far from Pasadena, Einstein and Elsa were Charlie Chaplin's guests at a dinner party that included Universal Studios founder Carl Lämmle, the actress Claudette Colbert, and the newspaper magnate William Randolph Hearst. Einstein and Elsa also joined Chaplin at the glittering premiere of his latest film, *City Lights*, and they joined Lämmle at a special showing of *his* latest film, *All Quiet on the Western Front*.

On their return journey in March, Einstein and Elsa traveled to New York by train, boarded the SS *Deutschland*, and arrived in Cuxhaven, at the mouth of the Elbe River. Seeing the throng of reporters and photographers waiting for him on the pier, Einstein asked the captain to let him and Elsa remain on board until the ship reached Hamburg, where they took a train home to Berlin.

Einstein and Elsa spent two more winter sessions at Caltech, in 1932 and 1933, and these turned out to be even more agreeable than the first, because there were many fewer publicity demands on him.

After his return from America in 1932, Einstein remained in Berlin for only six weeks: He had been invited to present the prestigious Rhodes Memorial Lecture in Oxford, and once there Einstein chose to linger for four weeks in his favorite academic refuge, Christ Church College. He took long walks in the lovely countryside around Oxford and met with scholars, scientists, and various student groups. He also took part in Oxford's active social life and made good use of the town's copious opportunities for playing chamber music.[17]

Two Weeks at Home

Einstein normally kept his travel diaries only while away from home. During his last two weeks in Berlin before leaving for Oxford, he made an exception—for reasons of his own—and kept a record of his activities while at home. Thanks to that fortunate circumstance, we have an account of Einstein's day-to-day routine in Berlin; it complements, as it were, the recollections of Fräulein Herta, who was his housekeeper at that time.

We learn from the diary that April 8, 1931, was a fine spring day in Berlin, and that Einstein passed the morning in his tower study. There he composed an essay on James Clerk Maxwell's conception of reality, in celebration of the centenary of Maxwell's birth—Maxwell was one of Einstein's greatest heroes. He then turned to resuming his ultimately fruitless search for a unified field theory.

The next morning Einstein met with Max von Laue and Erwin Schrödinger, two of his closest colleagues and friends, and traveled to Potsdam with them.[18] The three professors—Einstein refers to them as the "peace doves"—were on a mission to resolve the latest dispute between the physicist Erwin Freundlich and the astronomer Hans Ludendorff, the director of the Potsdam observatory.[19] Freundlich made use of the Einstein Tower's solar telescope to investigate whether light from the sun was "red-shifted" by the sun's gravitational field, as predicted by the general relativity theory. Although the Einstein Tower was part of the Potsdam observatory, it enjoyed a measure of autonomy because it was privately funded—an arrangement that led to frequent clashes between the high-handed Freundlich and the autocratic Ludendorff, but on this occasion, the "peace doves" succeeded in calming the waters. Einstein then returned to Berlin and spent an enjoyable evening in the company of Franz Oppenheimer, a noted economist and sociologist, much admired by Einstein for his brilliance and for the superb jokes he told—which Einstein, unfortunately, failed to record in his diary.

The unified field theory continued to occupy Einstein on a daily basis, sometimes together with his "calculator" Dr. Mayer. On April 11 he called on Professor Wolfgang Windelband at the Prussian Ministry of Education to plead (unsuccessfully) for an academic appointment for Mayer. Later, in the afternoon, he had a chat about America with a young nephew of Oppenheimer, who planned to emigrate to the United States to find work there as a gardener.

The following morning three musical acquaintances, all professional musicians, came to the apartment and joined Einstein in playing string quartets by Mozart, Schubert, and Brahms. In the afternoon, he paid a visit to Erich Mendelsohn, the architect of the Einstein Tower, and his evening was again devoted to chamber music: he played piano trios by Haydn, most likely with the pianist Joseph Schwarz and his son, Boris, Einstein's musical companions over many years.[20] After these two left, Einstein climbed up to his tower study and began an article on general relativity and the cosmological problem. This is the paper in which Einstein formally abandoned the cosmological constant in view of convincing evidence that the universe was expanding—evidence that Hubble had shown him a few weeks earlier.

Next day, Helen Dukas came to work with Einstein on his voluminous correspondence. Among it was a letter addressed to President Masaryk of Czechoslovakia on behalf of a young man who had refused military service—a stand that Einstein advocated staunchly at that time. After Dukas left, he completed the cosmological paper he had begun the night before. He submitted the manuscript in person at the Academy on April 16 and stayed on to attend the weekly Academy meeting. On that day Einstein found himself intensely irritated by the inanity of a lecture by a philologist who discussed certain expenditures of the Emperor Augustus. He went to the theater in the evening—we don't know with whom—and he thoroughly enjoyed the performance of the celebrated actress Käthe Dorsch in Lessing's comedy *Minna von Barnhelm*.

Einstein maintained a similar routine in the next few days: He worked on an article that listed field equations that were appropriate for a unified field theory and asked Mayer to proofread it. Together with Elsa he attended dinner parties where the guests included academics, bankers, diplomats, businessmen, artists, and even the police commissioner of Berlin. He also had lunch with the banker Henry Goldman, with whom he formed a friendship that would continue after Einstein settled in America.[21] He devoted one afternoon to selecting a piano for his sister, Maja, finally settling on a Blüthner instrument that was duly shipped to her in Italy. Back at home, he composed a speech in defense of the mathematician Emil Gumbel in Munich. Gumbel had written a pamphlet listing hundreds of political murders committed by the Nazis, and had become a special target of Rightists at his university.

On April 19, he went to the theater at the invitation of Estella Katzenellenbogen, the wealthy owner of a number of flower shops in Berlin, who often put her limousine and chauffeur at Einstein's disposal. They saw *The Blue Boll*, an avant-garde drama by Ernst Barlach, a playwright and sculptor best known for his antiwar activism—after he had enthusiastically volunteered to fight at the beginning of WWI. Two days later Einstein spent "noon and afternoon" with Toni Mendel and discussed Freud's recently published *Civilization and Its Discontents* with her. Although Einstein admired Freud's writings, he maintained a skeptical attitude toward psychoanalysis.

Two days later, at the next weekly Academy meeting, Einstein heard a lecture about catalysis by Fritz Haber, before attending a faculty meeting of the Academy. Afterward, he visited the home of a wealthy Berlin merchant where he heard a "wonderful" performance of negro spirituals. He then paid a visit to an old acquaintance, the blind and ailing writer Alexander Moszkowski, the author of the first Einstein biography.

April 29 was Einstein's last day in Berlin, and he spent it working on his correspondence with Helen Dukas and then took the evening train to Hamburg. He spent the night in a hotel, and in the morning he boarded the luxury ocean liner *Albert Ballin*, which would bring him to Southampton. The ship's name may well have brought a wry smile to Einstein's lips, for it was named after the former managing director of the Hapag Shipping Line. Albert Ballin had been one of Kaiser Wilhelm's most devoted *Kaiserjuden*, and in 1918, when the war was lost and the Kaiser fled to Holland, Ballin had committed suicide.

Weimar Culture and Science

Born in the aftermath of war, amid revolution and violence, the Weimar Republic represents Germany's faltering experiment in parliamentary democracy. The republic managed to survive its birth pangs, and it even flourished briefly in the mid-1920s, before the great depression brought it to its knees and the Nazis put an end to it. In its dozen years of existence, it nevertheless gave rise to a new awareness of cultural freedom and to an exhilarating creativity that extended from the arts, architecture, and design to music, film, and science.

Indeed, as gray as was its politics, so brilliant was the cultural blossoming in the Weimar Republic. Jazz music and new American dances such as the Charleston swept the country, and Walter Gropius's Bauhaus movement, along with Ludwig Mies van der Rohe, Erich Mendelsohn (Einstein Tower), and others introduced a new aesthetic in architecture and design. Artists like Käthe Kollwitz, George Grosz, and Emil Nolde portrayed society with a critical eye, and writers like Bertolt Brecht, Thomas Mann, and Carl Zuckmayer drove literature into new directions. In the hands of Fritz Lang (*Metropolis*, *Dr. Mabuse*), Friedrich Murnau (*Nosferatu*), and Josef von Sternberg (*The Blue Angel*), the new medium of film became an art form that was accessible to the masses. In music, Bertolt Brecht and Kurt Weill gave opera a brand new voice (*Three Penny Opera*, *Mahagonny*), and in medicine Sigmund Freud invented psychoanalysis, which soon became a part of popular culture.

Whether or not by coincidence, a similar blossoming took place in science, and particularly in physics. Ever since 1905, when Einstein invoked the existence of photons to explain the photoelectric effect, he had urged physicists to search for a new mechanics that was able to deal with quantum phenomena (e.g., with photons and electrons), but his counsel bore no fruit for many years.

In 1922 Arthur Compton performed an experiment that endowed photons with ever-greater reality by showing that their frequency changed after being scattered by electrons; that is to say, that they acted like particles, with energy and momentum being conserved. Niels Bohr created the first crude shell model of the atom that accounted for the chemical properties of the elements of the periodic table.[22] James Chadwick discovered the neutron, Wolfgang Pauli refined Bohr's model, Louis de Broglie proposed particle waves, and a veritable cascade of fundamental discoveries yielded new insights into the quantum world.

Then along came quantum mechanics, discovered independently by Werner Heisenberg and Erwin Schrödinger. It provided a mathematical formalism for dealing with phenomena on the atomic scale, and it was only fitting that the two men were nominated for the Nobel Prize by Einstein—arguably the "godfather" of quantum physics.

Following Heisenberg's uncertainty relationships (1926), most physicists accepted the "Copenhagen convention," espoused by Bohr, which interprets quantum mechanical results in terms of probabilities. Einstein, however, dissented from it, and this led to endless disputes with Bohr. Einstein, no longer the rebel of old, was unwilling to accept that "God played dice"—to the great distress of his colleagues and admirers.

From here on, Einstein retreated more and more from the mainstream of physics and pursued his solitary quest for a unified field theory. Having long maintained that "the Lord is subtle, but not malicious," he mellowed sufficiently with age to grant to his friend Hermann Weyl in 1951 that "maybe, he *is* a little malicious, after all."[23]

PRINCETON (1933–1955)

Farewell to Europe.

As Einstein's third and last winter session at Caltech came to a close, his plans for the return journey home were seriously disrupted by the dramatic events taking place in Germany. On January 30, 1933, the aging President Hindenburg charged Adolf Hitler with forming a government made up of members of monarchist and Nazi parties. Within days, Hitler used an "emergency edict for the protection of the German *Volk*" to outlaw all opposition parties, dissolve the Reichstag, and abolish press freedom. The Third Reich had begun.

On March 10, the day before he left Pasadena, Einstein announced at a news conference that he would not return to Germany and would, from now on, only live in a country that had "political freedom, tolerance, and equality before the law for all citizens." He also confirmed that beginning in March 1934 he would be associated with a new research institution, the Institute for Advanced Study in Princeton.

Einstein and Elsa traveled by train to New York, where Einstein was prevailed upon to take part in a Zionist rally and also in an antiwar event. Instead of returning to Germany on the *Deutschland*, as originally planned, Einstein booked passage to Antwerp on the *Belgenland.* Immediately upon landing, he traveled by taxi to Brussels, where he called on the German consulate to renounce his German citizenship and to hand in his passport. His letter of resignation to the Prussian Academy of Sciences stated that he found it intolerable to serve the Academy under the prevailing circumstances, while it also expressed his gratitude to the Academy for many years of intellectual stimulation and for his personal relationships with its members. Max von Laue recounted later that the letter infuriated the ministry officials, who had already ordered the Academy to prepare formal disciplinary proceedings against Einstein.

Einstein was now unemployed, but not destitute, thanks to his hard currency savings. He rented a modest cottage in the small Belgian seaside town of Le Coq sur Mer, where he was observed taking solitary walks along the dunes—instead of in the forest above his country home in Caputh. In the past, Einstein had vigorously promoted a militant pacifism, but he recognized that once Hitler had attained power in Germany his position had become naïve and untenable. The issue came to a head when Belgium's King Albert asked Einstein to comment on the case of two draft resisters in a Belgian jail. Einstein, who had enjoyed warm relations with the royal couple in calmer days, when he played chamber music with Queen Elisabeth, responded that under the changed circumstances, he had no qualms about changing his views.[24] In a letter that he allowed to be published, he stated that Germany had created a situation in which the Belgian army was urgently needed for defense and that he, too, would not refuse military service, indeed, he would gladly accept it "in the knowledge that he would be serving European civilization." The letter made headlines around the world and stunned Einstein's many followers, and particularly the War Resisters' International. But he remained firm.

During his stay in Le Coq sur Mer, Einstein traveled to Zurich to see Eduard and Mileva, in what turned out to be their last meeting. Einstein had brought his violin and played with "Tetel" as of old, but the visit was a wrenching experience for both parents. From Zurich, Einstein went to Britain, where he lectured in Oxford and Glasgow. He made a second trip to England as the guest of Oliver Locker-Lampson, a Member of Parliament with a colorful past, and stayed at his host's secluded Norfolk home. Locker-Lampson took Einstein to the House of Commons and also to Chartwell for a lunch with Winston Churchill, whose political career was then at a low ebb.[25]

In October, his host drove Einstein to Southampton, where he joined Elsa, Dukas, and Mayer, who were already on board the *Westmoreland*, the ship that would bring the four newest refugees to America. He did not know it then, but it was the last time he would see Europe.

Coda

When the *Westmoreland* reached New York harbor, a festive welcome was prepared for Einstein at the Twenty-Third Street pier, but the ship arrived without its famous passenger. Abraham Flexner, the director of the brand-new Institute for Advanced Study, had arranged for the Einstein party to be brought ashore in a tugboat before the ship docked, in accord with his dogged policy of shielding Einstein from all publicity because he was afraid that it would create an anti-Semitic backlash. Once ashore, the party was whisked off to Princeton, where rooms had been prepared for them at the Peacock Inn, and a few days later the party moved into an apartment near the Theological Seminary. It was the Einsteins' first real home since leaving Berlin for America almost a year before. Einstein's office was temporarily located in the Mathematics Department of Princeton University since the Institute's facilities were not finished.

Flexner's overprotective policies came to a head when he intercepted an invitation to dine at the White House, which was addressed to Einstein. When Einstein discovered this, he was furious and threatened to resign from the Institute. Flexner gave in, but as a consequence of this brouhaha Einstein lost all influence over the Institute's future policies. On January 24, 1934, Einstein and Elsa did dine with the Roosevelts at the White House and spent the night there. Though the plight of Germany's Jews must have been uppermost on Einstein's mind, the president preferred less political topics of conversation, such as sailing.

Einstein practiced remarkable restraint in discussing political issues in public. He understood that the American public would probably not be receptive to the views of a newcomer who had himself been victimized by the Nazis. What distressed him most was his conviction that Hitler was heading for war, a prospect that the democracies were not yet prepared to face.

In May 1934 the Einstein family received the devastating news that Ilse was dangerously ill. Elsa traveled to Paris to be with her, but could only watch her die. Elsa was desolated. Einstein, who had by now settled in his new home, had no wish to accompany Elsa and to return to Europe. He passed the summer day-sailing along the coast of Connecticut, as a guest of Gustav Bucky, his physician and friend since their days in Berlin. Later that year, Margot and her husband, Dmitri, arrived in Princeton, although they divorced soon afterward.

In 1935 the white clapboard house at 112 Mercer Street, in the same neighborhood as the Einsteins' rented apartment, came on the market, and Einstein bought it for cash. There was enough money left over for some renovations, and Elsa had a study with a large picture window constructed on the upper floor—an evocation of Einstein's tower study in Berlin. The remainder of the house was furnished with what was rescued from the apartment in the Haberlandstrasse, including the grand piano that once stood in the Biedermeier room. Sadly, Elsa had little time to enjoy their new, and also their last, homestead. She fell ill with kidney disease and, after a painful summer and fall, she died in December 1936. Her passing hit Einstein much harder than he might have expected: their romance had evolved into a friendship in which each understood and respected the other's needs.

In 1937, Hans Albert and his family also arrived in America, and two years later, shortly before the Second World War broke out, Einstein's newly divorced sister, Maja, came as well. When she joined Margot and Helen Dukas at 112 Mercer Street, Einstein had a trio of utterly devoted housemates to take care of him.[26] He treasured his tranquil existence as he continued to pursue his scientific and humanitarian passions. To indulge in his favorite pastime, he bought a seventeen-foot sailboat, and in homage to the lost *Tümmler* he named it *Tinnef*—Yiddish for "a piece of junk."

In the 1930s, Einstein still made occasional sallies against Bohr's views on quantum mechanics, but they all missed their mark. He became increasingly isolated from contemporary physics as he continued his search for a unified field theory. When a colleague asked him why he persisted in his lonely quest, Einstein replied that the chance of success was small, but the attempt was worthy.

Yet Einstein's name was still one to be reckoned with. In the summer of 1939, when war was on the horizon, he was vacationing on Long Island when he received a visit from the eminent theoretical physicists Eugene Wigner and Leó Szilárd. They informed him that recent nuclear physics experiments had shown that neutrons created during fission of uranium nuclei could initiate a chain reaction with the release of a vast amount of energy—in accordance with Einstein's formula $E = mc^2$. The two visitors feared that German scientists might be working on harnessing this energy for military purposes and urged Einstein to notify President Roosevelt of this potentially dire development. Einstein readily agreed and signed such a letter to the president, a letter that presaged the Manhattan project by two years and may have played a role in making the government aware of the potential for an atomic bomb. Einstein commented later that he would not have sent that letter had he known that the fears of Germany developing a fission bomb had been groundless.

Also in 1939, Niels Bohr spent several months at the Institute in Princeton, and a poignant meeting of these two giants of twentieth-century physics took place: The two men had loved and admired each other twenty years earlier, but their long dispute had taken its toll, and now they did little more than exchange cursory greetings.

Princeton's most famous citizen was best known as an absent-minded professor who sometimes lost his way on his walks along suburban streets and for helping the neighborhood children with their homework.

In 1940 Einstein became a US citizen. At the end of the war, Einstein, horrified by the nightmare just past, was determined to sever almost all his links to his former "step-fatherland." But this clearly did not apply to Fräulein Herta, who still resided in Berlin, now the capital of communist East Germany. In 1947 she sent him birthday greetings and received the following response from Einstein—which was followed by several, very welcome food packages:

> Dear Herta!
>
> You should have seen the jubilation when your letter arrived. How often had we asked ourselves what might have happened to our dear courageous Herta? And we were concerned that we heard nothing for such a long time. After all, you lived through so many different experiences together with us, in the now vanished Haberlandstrasse and in Caputh. The Russians now have the little house and a Nazi got the plump sailboat. I hope that things will soon improve for you with regard to the wretched food supply.
>
> Many warm good wishes,
>
> Your
>
> A. E.

Einstein resigned from the Institute in 1944, but his daily routine changed little. He continued his walks along Princeton's suburban streets, alone or in the company of Kurt Gödel, Wolfgang Pauli, or another German-speaking associate, for he never learned to feel at home with the English language. After Hiroshima, he was a passionate supporter of nuclear disarmament, and in the early 1950s he was a vocal opponent of the McCarthy witch hunt—stands that earned him kudos and warm greetings from his old friend Toni Mendel who had settled in Canada.[27]

In 1948 Einstein suffered painful abdominal attacks that were ascribed to a grapefruit-sized aneurism in his abdominal aorta. His sister, Maja, meanwhile, suffered a stroke and was bedridden for many months, during which time Einstein read Cervantes's *Don Quixote* to her, as well as many other books. She died in 1951.

Four years later it was Einstein's turn. The aneurism perforated and he was admitted to the hospital, where he rejected all measures to extend his life as "tasteless" [*geschmacklos*]. He died five days later, one month after his seventy-sixth birthday.

Einstein's stepdaughter Margot visited him in the hospital shortly before he died, and in a letter to Hedwig Born she wrote: "He spoke with profound serenity—even with a touch of humor—about the doctors, and awaited his end as an imminent natural phenomenon. As fearless as he had been all his life, so he faced death, humbly and quietly. He left this world without sentimentality or regrets."[28]

CHAPTER 4
EINSTEIN AT HOME
Herta W. Recalls the Years 1927 to 1933

by Friedrich Herneck

Translated and Annotated by Josef Eisinger

FIRST CONVERSATION: THE APARTMENT IN THE HABERLANDSTRASSE

Friedrich Herneck: You were kind enough to make yourself available for answering questions about Einstein's final years in Berlin, and I am grateful to you. We agreed that I would participate in our conversation so that I can vouch that the information you provide is truly yours. I beg you to excuse me if, unavoidably, my questions may at times feel like an interrogation. I ask you to answer them, all the same, and as best you can.

Herta Waldow: I will do my best.

FH: We realize that after the passage of some fifty years the memories of some events may have faded or, occasionally, vanished. We must therefore be content with fragmentary reminiscences and confine ourselves to events and observations that, for whatever reason, are firmly anchored in your memory.

I begin by asking you for information about yourself.

HW: I was born on December 20, 1906 in Salzhof near Spandau. My father, whose parents had emigrated to America with six of their eight children, was employed as an "executive coachman" at a chemical plant. As coachman for the plant's directors, he also drove their children to school.

My mother was the daughter of a miner from Upper Silesia, and before she married she worked as a domestic servant on an estate in Westhavelland, and as an agricultural worker. At the end of the First World War, the factory that had manufactured war materials closed, and my parents moved to Lautawerk in Niederlausitz. My father had been recommended for a job at the "United Aluminum Works" in that town.

FH: According to the police files that brought you to my attention, the police looked for you at your parents' home in Lautawerk in the summer of 1934. The Gestapo had instigated the criminal investigation police to interrogate you about Einstein.

HW: Yes, there was great commotion in town when the police looked for me. But I had long since taken another position in Berlin, in the household of a doctor.

FH: Did you have any siblings?

HW: Two brothers and a sister.

FH: Where did you go to school?

HW: I attended public school in Spandau from 1913 to 1920, and from May 1920 till the end of the school year in Lautawerk. In Spandau I had been selected to attend an upper school, but unfortunately nothing came of it because of my parents' relocation.

FH: Do you know anything of your father's political views?

HW: No, I only know that he took part in the May Day celebrations of the Social Democrats in Spandau.

FH: According to the police transcript of your interrogation of September 5, 1934, you were with Einstein, as a domestic servant, from June 15, 1927, until June 1, 1933. The testimonial Frau Elsa [Einstein] wrote for you in May 1933 mentions a somewhat earlier time for the start of your employment.

HW: I believe the data I gave at my hearing correspond to the facts.

FH: Where did you work before you came to Einstein?

HW: My father did not want me to take an office job, although that was my wish. After leaving school, I was therefore obliged to spend a year cleaning rooms in a home for bachelors in Lautawerk. Then I accepted a position in Berlin with Geheimrat H., who was a member of the board of the United Aluminum Works, and before long I had to run the entire household there. In the winter of 1926/27 I had an appendectomy operation in the Charité [hospital], and since I had a difficult recovery Frau Geheimrat thought it best that I should, temporarily, look for a less strenuous job; and so I took an easy job in a soap factory in Berlin, and then I came to the home of Professor Einstein.

FH: Through a recommendation?

HW: No, I had gone to an employment agency—I think it was in the Jägerstrasse—and Frau Einstein came there at the same time and saw me. We chatted for a while and then she took me right away to the Haberlandstrasse, to show me the apartment. I accepted her offer and started work immediately.

FH: Frau Einstein was evidently very eager to have you start right away.

HW: Yes, the Einsteins were under a lot of stress because the previous domestic servant had left quite precipitately, taking several valuables with her. Sometime later, I went to the courthouse with Frau Einstein and we brought back the stolen articles she had taken. They were silver objects and, well, whatever else she considered worth taking along. As a result, the Einsteins were utterly alone and were glad to have found someone so quickly. I was twenty years old at the time.

FH: Did Frau Einstein inquire about your housekeeping experience?

HW: She asked if I knew how to cook. I said I knew a little about cooking but I was no cook. And straight away I cooked their midday meal of lamb chops and string beans. Afterward, Herr Professor said that it was the first time somebody said she didn't know how to cook and then prepared something that tasted better than anything he had ever eaten. I can recall that very distinctly, as if it happened only yesterday.

FH: Well it was high praise for you, after all. Did you know anything about Einstein before you began?

HW: Not much. I had, of course, heard his name and may have read something in the newspaper, but it had not left an impression on me.

FH: What was your impression of their building? Judging from old photos, it was built shortly before the turn of the century.

HW: From my standpoint, it was somewhat ostentatious. It had an elevator and a doorman who sat downstairs in the porter's lodge, which was connected to his apartment.

FH: On which floor were the Einsteins?

HW: On the fourth floor. It was a seven-room apartment. One floor up, in the attic, was the study of Herr Professor.

FH: You told me at one time that directly below them, on the third floor, lived the SPD [Social Democratic Party] Reichstag delegate Dr. Rudolf Breitscheid—who in 1944 lost his life in the Buchenwald concentration camp. Did the two families have any social contacts?

HW: The Breitscheids had moved in only quite recently, toward the end of 1932, after Herr Professor had already left for America to lecture again in Pasadena. They may have met on other occasions; I don't know about that. But in the building they had no contact.

FH: Who were the other tenants in the building?

HW: I cannot tell. Each tenant had a floor to himself, alone. There were no friendly contacts among the tenants, nor among the domestics. As a result, I don't know who lived there, other than the Breitscheids.

Visitors to Einstein's apartment have expressed diverse opinions about it. Max Born, in his reminiscences of Einstein, spoke of a "cultivated domesticity." Philipp Frank—Einstein's successor as professor of theoretical physics at the German University in Prague—wrote in his biography: Einstein "lived amidst beautiful furniture, carpets and pictures" and considered the apartment to be "spacious."[1] Frank had got to know the apartment already during the First World War, when Einstein—then still living in the Wittelsbacherstrasse—had taken him there for dinner, when it was occupied by his uncle-twice-removed, Rudolf Einstein. Writing to his friend Besso in Bern, Einstein also called the apartment "spacious." Charlie Chaplin, on the other hand, who had visited the apartment in the Haberlandstrasse in 1931, wrote in his book *The Story of My Life* that Einstein lived in a "modest, small apartment"' with "old, well-worn carpets." Seen from the point of view of an American multimillionaire, used to correspondingly comfortable accommodations, Einstein's seven-room apartment may appear so. But for German circumstances, the apartment was neither small nor modest.

FH: I would now like to ask you to describe the location of the rooms, based on your thorough knowledge.

HW: Viewed from the apartment entry, all the way in the back on the left side was the bedroom of Ilse and Margot, the two daughters from Frau Elsa's first marriage. When I first arrived, it was Margot's room, because Ilse had very recently married and was no longer living in the Haberlandstrasse. Just in front of that room was the living room [*Salon*], the so-called Biedermeier room.

FH: Forgive the interruption, but the word Biedermeier reminds this aficionado of Einstein's life of his school days in Munich. At that time, his class mates called him "the Biedermeier" in good-natured banter because he spoke so slowly and deliberately and was so upright and staid.[2] But please continue to describe the apartment.

HW: Next to the living room was the library, and adjoining the library was Herr Professor's bedroom, right next to the apartment entry. On the right side, all the way in the back, was the bedroom of Frau Professor and, next to it, the bathroom. Then came the dining room, with a door leading to the Biedermeier room. Then came a little hallway that led to the guest room where visitors occasionally spent the night. But this did not happen often. Most of the time, the room was used for sewing and ironing. Then came the kitchen on the right side, and next to it the pantry with a large refrigerator, which was cooled with large blocks of ice. In Caputh, we had an electric refrigerator, which, I hear, is still there . . .

FH: . . . but does not work anymore. That would really be asking too much after almost half a century!

HW: At the very front, right next to the entrance, was the toilet. I, myself, had a small room on the far right side. The largest room of the apartment was the living room, where the grand piano stood and Herr Professor's violin case usually sat on it. The piano did not really fit in with the Biedermeier decor, in my opinion, but it was there because it was the largest room in the apartment.

FH: Did the piano have a beautiful sound? It was supposedly a Bechstein grand.

HW: I cannot recall the make, but it was a very fine instrument, in my view.

FH: In which direction did the windows of the Biedermeier room and the library face?

HW: I suppose east or southeast, for these rooms were always very bright.

FH: Southeast would agree with the layout of the streets that I recently saw in an old city map of Berlin. Unfortunately, this cannot be confirmed visually; as you know, the building on the Haberlandstrasse was destroyed right down to its foundation during an air raid in the Second World War.

HW: I visited that neighborhood once after the war. It was unrecognizable.

FH: The house in Ulm, where Einstein was born, also fell victim to an air attack and was utterly destroyed. Fortunately, the summer house in Caputh survived the war intact—contrary to some statements in the literature—although a landmine landed nearby, during a raid on Potsdam in April 1945. It caused a lot of damage and took the lives of several inhabitants. In all the territory of the two Germanies, it is today the only house that stands as a reminder of Einstein. But let us return to the apartment on the Haberlandstrasse, circa 1930. An attic room, one floor up, had supposedly been constructed for Einstein in the 1920s. Could you describe this room, his actual study, in some detail?

HW: The tower room [*Turmzimmer*] as we called it, consisted originally of two attic rooms, and during the construction the wall between them was partly removed. The room was wide, but not deep, and one wall was filled with books.

FH: Two small windows can be seen in the photo.

HW: Yes, you faced two windows as you entered the room. In one corner, the floor was raised slightly, and that is where Herr Professor always sat. His desk was there, and this is where he worked. Next to the desk was his telescope for observing the stars. It was a kind of school telescope mounted on a tripod and it stood in the corner, next to the window.

FH: Even an amateur telescope lets one observe the craters of the moon, Galileo's moons of Jupiter, the rings of Saturn, the phases of Venus, and much else. Did Einstein ever invite you to look through the telescope?

HW: Sadly, no.

FH: Dr. Rudolf Kayser, the husband of Einstein's stepdaughter Ilse, published an Einstein biography in English in 1930, using the pseudonym Anton Reiser. Einstein wrote a preface for it. Kayser who knew the layout in the Haberlandstrasse very well, wrote that on the walls of Einstein's study hung three pictures. They were the likenesses of the English physicists Michael Faraday and James Clerk Maxwell, and of the German philosopher Arthur Schopenhauer, whose writings Einstein liked to read. Other biographers mention a picture of Isaac Newton. Did the tower room have enough free wall space to accommodate these pictures?

HW: The wall next to the telescope was free, and so was the space between the two windows, where there were no bookcases, as well. It was therefore quite possible to hang three or four pictures there. It was a real attic room with slightly sloping walls in the front. Herr Professor would often retire to this room for several hours. His secretary—in 1927 it was a student—always used to go up there to do the typing.

FH: The student who came to the Haberlandstrasse to do Einstein's typing while you were there was registered in the law faculty of Berlin University. Before him, another student had typed for Einstein, and he, too, was a law student. That is not a coincidence, for at that time law students often used to skip their professors' lectures because they could get the examination material just as easily from books. They had more time for outside work than, say, a science student who had experimental work to do. By the way, Einstein's first private secretary was his stepdaughter Ilse, who did his typing until she married in 1926.

HW: That was before my time.

FH: In the spring of 1928, when Einstein was sick in bed, Fräulein Dukas—as you know—came into the house as his private secretary, initially on a trial basis. She stayed on, however, for over a quarter of a century, and since 1955 she has been the administrator of Einstein's literary estate in Princeton.

HW: I remember Fräulein Dukas very well, although I had no particular dealings with her. That's because she did not live in the Haberlandstrasse but came a few times a week, maybe three times, for dictation and to take care of the correspondence. She did the typing upstairs in the tower room—on a portable typewriter.

FH: You have mentioned to me that Einstein was once painted in his study. Can you tell me more details about that?

HW: Yes, Herr Professor was painted in the tower room, and when I came upstairs to bring tea or other refreshments I had to confirm that the painting was beginning to resemble him. The painter asked me to do so, and Herr Professor appreciated my judgment. I had to validate the resemblance as a lay person, so to speak.

FH: There are several paintings of Einstein that date from his Berlin period; one of them was painted by Max Liebermann.

HW: The painter I am speaking of was certainly not such a famous artist as Professor Liebermann. I would have recognized him, in any case, from his photos; and he would surely not have asked me if the portrait he was painting caught his likeness.

FH: Liebermann painted his portraits generally in his studio. The famous portrait of Ferdinand Sauerbruch that stems from the time you were with Einstein was also painted in Liebermann's studio on the Pariser Platz. The famous surgeon had to bring his white surgical gown with him. The painter you mentioned must have been a little-known artist.

HW: It seemed to me that the artist was somebody to whom Herr Professor wanted to give a chance of becoming publicly known through his portrait. In the beginning, he spent a lot of time observing Herr Professor while he worked, in order to memorize his facial features. Then he set up his easel and actually started to paint. And I had to look at it frequently.

FH: It is typical of Einstein's democratic mindset that he should place such a high value on the judgment of a layperson, which you surely are with regard to painting. This is, for me, an expression of the respect Einstein always accorded to so-called ordinary people. This is a trait one encounters on many other occasions.

Earlier, I mentioned the contradictory opinions about Einstein's home decor. What was your impression of the apartment?

HW: Well, now, the furnishings were all very fine and beautiful. The sideboard in the dining room filled the entire wall space, and all the furnishings were very solid. The Biedermeier room was, from my standpoint, how shall I put it, very tasteful. I have not exactly been spoiled in such matters, nor am I an expert, but I must say that I found it all genuinely beautiful. I was particularly impressed by the library, with its great wall of books. The books reached all the way to the ceiling. I had never seen anything like this before, not even in the home of Geheimrat H.

FH: If you add to that the crowded bookshelves in the tower room, one can hardly claim that the Einsteins had books, but not a library. That is, however, what János Plesch, of whom we will speak more later, says in his 1949 memoir, *A Physician Recounts His Life.*[3]

HW: I don't understand how Professor Plesch could make such an assertion. He was quite familiar with the apartment, after all.

FH: Were there only literary works in the library, such as novels, stories, poetry, etc., or specialized books?

HW: There was only leisure time literature. The scientific works that Herr Professor needed for his work were all upstairs in the tower room.

FH: Were there any larger, multivolume reference works among them, such as the "Brockhaus" or "Meyer" [encyclopedia]?—because Plesch claims that Einstein did not possess any reference works.

HW: That might well be right. I cannot recall any multivolume reference works. I would surely have noticed them, for I dusted the books in the tower room regularly; once a week, when Herr Professor was absent.

FH: What else was in the library, other than books?

HW: A narrow table, a few chairs, and a comfortable armchair. There was a desk at the window and, on it, a large globe. Upon entering, on the left, there was a glassed case with books inside, the only glassed-in bookcase in the apartment. Apart from that, I can only recall the shelves of books.

FH: You told me that there was a balcony adjoining the living room. Was it used often?

HW: Only rarely. I also cannot recall if there were any plants. There was hardly anything green in the apartment . . .

FH: . . . Apart from the small Norfolk pine, an araucaria, that can be seen in the photo of the living room in front of the window on the right.

HW: I know nothing of a little tree like that. The photo may have been taken before my time.

FH: How was Einstein's bedroom furnished?

HW: Very simply. His bed stood against the library wall, with a night table beside it. There was also a wardrobe, a table, and a few chairs. Next to the door was a cabinet with the silver cutlery for twenty-four persons, including fish cutlery, mocha spoons, and everything else that belongs to it. The room was kept very plain.

FH: From some memoirs of Einstein's Berlin years it appears that when strangers wished to visit the scientist in the Haberlandstrasse, they usually called first on the telephone. Where was the telephone?

HW: The main connection was in the short passageway that led to the guest room and the kitchen. There were several secondary outlets that I could connect to, depending on whether the call was for Herr or Frau Professor. Herr Professor had two handsets, one downstairs on his night table and one upstairs on his desk next to the window. That was also the one that was used to call him to dinner. When a call came in, I always answered with: "Here, at Professor Einstein's." Most of the telephone calls were taken by Frau Professor. He used the telephone only for scientific or very private conversations. Everything else was taken care of by Frau Professor. But if, for example, Geheimrat Planck or other scientists called, I had to connect to Herr Professor's room in the downstairs apartment or upstairs in the tower room.[4] It was an old-fashioned instrument of the kind still in use at the time: The box was mounted on the wall, with the receiver hanging alongside. When you wanted to make a call, you first turned a little crank and waited for the telephone exchange to answer. Direct dialing like we use today did not exist then. All Berlin was divided into telephone districts, whose exchanges had specific names. I cannot recall the name of our exchange.

FH: I checked the Berlin telephone directory from 1931, and Einstein's number was 2807 and the exchange name was Cornelius. Were there many calls?

HW: Yes, indeed! Some days there were a great many. Invitations, in particular, were arranged by telephone. Margot also made a lot of calls. The elder stepdaughter, Ilse, was already married in my time and lived elsewhere. Her husband, whom you mentioned as the author of an Einstein biography, was an editor with the S. Fischer publishing firm. A book by him appeared there during my time.

FH: It was probably his book about the French mathematician Henri Beyle, who used the name Stendhal as a novelist.[5] He was one of the great realist writers of the nineteenth century. His novels *The Red and the Black* and *The Charterhouse of Parma* became widely known after 1945 when they were made into films. I read Kayser's book some time ago. It is an intelligent, engagingly written biographical work. It contains descriptive passages of Stendhal that are reminiscent of Einstein, for example: "He wore his hair long; he did not want to wrest from his mathematics the half-hour a haircut would cost him."—That is something that might also have been said about Einstein. The book carries the dedication "To my Ilse." Dr. Kayser apparently loved his wife very much.

HW: The Kaysers had a very happy marriage. They often came to visit in the Haberlandstrasse, which was not far from where they lived. Frau Kayser was already quite poorly, and everyone was very concerned about her. Later, they both often came to Caputh, where a room was always kept ready for them.

FH: Since you brought up hair cutting: I recall hearing from you that Frau Elsa often cut her husband's hair herself, because she could not get him to pay a visit to the barber. This is also confirmed by his friend Plesch, who wrote of Einstein: "He does not spend a great deal of time on his adornment. Nobody can tell how long his tresses would have grown if Frau Elsa had not, from time to time, cut his hair with scissors."[6]—It seems that Frau Elsa often played the role of an amateur hairdresser.

HW: Yes, she did. When his hair was too long, when it was beyond the pale, she would cut off his hair with scissors, and he was willing to put up with it. Since Frau Professor was very nearsighted and unable to hold her lorgnette while cutting . . .

FH: . . . she was, so to speak, flying blind, and the haircut turned out accordingly. There exist photos in which Einstein's head looks like a plucked chicken. That might be the result of a session in "Salon Elsa Einstein."

HW: Yes, but Herr Professor just could not be moved to visit a professional barber.

FH: Plesch has written that Einstein's moustache was likewise "only trimmed in an amateurish manner," only when it bothered him too much. Judging by photos, this seems to be true, too. But let us return to the apartment in the Haberlandstrasse. Apart from the main entrance, it must have had a service entry for delivery people, messengers, tradesmen, beggars, etc., as was customary in bourgeois apartment buildings of the time.

HW: Yes, the landing of that staircase was right next to the kitchen. By using it, you emerged not on the Haberlandstrasse but in the Aschaffenburgerstrasse, at an appropriately marked entrance. I was not required to use it because, right from the start, I had a key to the building and to the elevator. The main entrance on the Haberlandstrasse was locked, day and night, and the porter opened it only for visitors of one of the tenants.

In April 1914, when Einstein moved with his family from Zurich to Berlin to take up his post at the Academy of Sciences, he rented an apartment in Dahlem, a rural suburb of the "Imperial capital." This is where, by 1911, the Kaiser Wilhelm Society for the Advancement of the Sciences had built a research institute for chemistry and where it was planned to build an institute for physics, with Einstein as its director. In the summer of 1914, Einstein's wife, Mileva, returned to Switzerland with their two sons, Hans Albert (ten) and Eduard (four), and once the war started they remained there. Since the planned physics institute was by then not going to be built, there was no longer any reason for Einstein to remain in a suburb that was somewhat inconvenient to reach. By 1915, he had therefore moved to Berlin-Wilmersdorf, Wittelsbacherstrasse 13, where, according to a visitor, he lived like a bachelor in an almost empty apartment. In September 1917 he moved to Haberlandstrasse 5, where he lived with his cousin Elsa, who had nursed him devotedly during his illness, and with her father, Rudolf Einstein, his future father-in-law. This was the third and last of Einstein's residences in Berlin.

FH: What was the nearest subway station?

HW: Bayrischer Platz. When the Einsteins wanted to go somewhere and a car was not at their disposal, they always used the subway or the bus, which also stopped on the Bayrischer Platz. It was the only large square in the neighborhood, and it had stores, bakeries, and pastry shops where I often went to shop.

FH: Was there a park with park benches near the apartment? The reason that I ask is that the Soviet physicist and Einstein's friend, Professor Abram Joffe, wrote in his memoir, *Encounters with Physicists*, that he once sat on a park bench near the Haberlandstrasse for quite a while because Einstein wanted to avoid a visitor whom Frau Elsa had invited against his wishes. Einstein brought his guest to the park to continue their scientific conversation without being disturbed. When the "danger" was over, the two physicists returned to Einstein's upstairs study and discussed problems of crystal physics until two in the morning. Joffe, a student of Röntgen, was deeply involved with these questions, while Einstein had a lively interest in the mechanical and electrical properties of crystals. That park must have been near the apartment. Do you recall such a park?

HW: I don't remember any large park nearby, but there were several small parks. Everything was still a bit informal in that neighborhood.

FH: Another question about the apartment. How was your room furnished?

HW: It had a large window, but by and large, it was a typical housemaid's room, as one said then. There was a bed, a washstand, a cupboard; but a table?—I don't think so, there wasn't enough room, and I would hardly have sat down at it. When Herr Professor was away and Margot was also not there, I had to sleep in the guest room, so that I would not be so far from Frau Professor's room. That is also where I slept when both Einsteins were away and only Margot was in the apartment—so that I would not be so far "from the action," so to speak.

FH: Were you able to hear the telephone ring from your room?

HW: Yes, yes. I could hear it easily. I could also hear it well when Herr Professor occasionally played the violin in the kitchen.

FH: We will turn to such occasional sleep interruptions later. First this: What were your tasks at the Einsteins?

HW: I acted as the housekeeper, *Stütze der Hausfrau*, as it was called. I had to run the household and do the cooking. There was no cook, but, when a larger company of fifteen or twenty persons was invited, someone came to the apartment to do the dishes.

FH: You can probably remember the menu at such parties. It may be interesting to hear from an eyewitness what delicacies were served on such occasions.

HW: There was always a clear bouillon with a savory egg custard garnish to start with, then salmon with mayonnaise, also with smoked salmon, followed by a pork filet with chestnuts, and afterward a strawberry dish with whipped cream mixed with strawberries to make "strawberry snow."

FH: Einstein seems to have been particularly fond of strawberries. Plesch has reported that when Einstein visited his country estate in Gatow, he ate strawberries by the pound.[7]

HW: Herr Professor was indeed passionately fond of strawberries.

FH: Was this always the menu at such large parties, or did it change from one time to the next?

HW: On greater occasions, the menu was, almost always, as I told you. Frau Professor always made the mayonnaise in the afternoon before—I mean the salmon mayonnaise—and helped with the strawberry dish. In the evening of the event I had to prepare the bouillon with the savory custard, the chestnuts, and the meat. I liked to cook . . .

FH: . . . and were good at it, no doubt, as is attested by Frau Einstein in her testimonial of May 1933. You did not have to do the heavy household work?

HW: No. That was always done by a cleaning lady who came from Schöneberg. She also washed the windows of the downstairs apartment. And when there was a lot of company, she helped with the preparation and the serving of the food and particularly, with washing the dishes. This lady did not come every day, however.

FH: Did such large dinner parties take place often?

HW: About three times a year. They were the obligatory responses to the invitations the Einsteins had received. Besides those, there were many afternoon teas and musical evenings in a smaller circle. Occasionally, when there was a presentation in the apartment, the circle was somewhat larger. The gathering was then more like a stand-up reception and I would go around passing out cups of tea. There were some small nesting tables that could be taken apart to provide a sufficient number of level surfaces that allowed guests to put down their tea cups.

FH: Did this take place in the dining room?

HW: In the three large rooms—the Biedermeier room, the dining room, and the library. These rooms were always used on such occasions.

HW: You told me once about presentations that were given in the apartment. Who gave the presentations?

HW: First of all, Herr Professor himself, for example, upon returning from a trip from America. Guests were invited, and Herr Professor talked about Pasadena and other cities he had visited on his voyage.

FH: You said earlier that you could hear Einstein's nighttime violin playing in your room. Didn't that bother you?

HW: Oh, not at all. Herr Professor played the violin at night only occasionally, not often.

FH: Why did he also play in the kitchen? There were plenty of other rooms.

HW: The walls in the kitchen were tiled, and the sound resonated nicely there. Herr Professor did not play pieces of music, but his own improvisations, and he did his thinking to them.

FH: I think this is a perceptive observation. Einstein did not make music to relax or to pass the time. Active music making was for him a component of the scientific thought process. Maybe one can put it this way: When he could not make progress with his formulas, he picked up his violin or sat down at the piano. Sometimes he then stopped abruptly and said: "That's it, now I have it!" [*So, jetzt hab ich's!*] That is told by his sister, Maja, and I consider her report to be trustworthy. It is reminiscent of the Greek mathematician Archimedes, who thought of the solution to a problem while sitting in the bathtub; whereupon he jumped out, shouting: "Eureka!" (I have found it!)

HW: Herr Professor was passionately fond of playing the violin and the piano.

FH: He apparently had no artistic inclinations apart from music—or did he perhaps draw or paint, as did the biologist Ernst Haeckel or the chemist Wilhelm Ostwald, and many physicians today—witness the current exhibitions "With Stethoscope and Palette"?

HW: I never saw anything like that with Herr Professor.

FH: It is also conceivable that Einstein was an amateur photographer. The 35 mm camera, embodied by the Leica, came on the market in 1925, and it gave photography a big push. Einstein had, moreover, undertaken many major journeys abroad that offered abundant subjects for photos. Did you ever see a camera?

HW: No.

FH: Was there a radio in the Haberlandstrasse?

HW: Yes. It stood on the window sill in the dining room, at the window that looked out on the courtyard. It was a rather large, black or dark-brown case with a separate loudspeaker sitting on top, as was usual in the early days of radio. There were no built-in speakers, as we know them today.

FH: They did not come into use until about 1930.

HW: I would like to add something regarding the dining room where the radio was: next to the large table in the center stood a smaller table where breakfast and supper were eaten. That was where the family ate every day.

FH: How many people could the large table accommodate?

HW: If table leaves were inserted, twenty to twenty-four. When needed, we fetched the table leaves from the attic storage room and inserted them to expand the table. Chairs were brought in from the other rooms and additional ones from the attic.

FH: Was there a gramophone?

HW: No, no, only a radio.

FH: That is understandable because János Plesch assures us that Einstein despised the "canned music" produced by records before the advent of electrical recording and reproduction techniques. Einstein was profoundly musical and had a sensitive ear, and the background noise associated with mechano-acoustical recordings bothered him. It was not till later, in Princeton, that he enjoyed works of classical music on long-playing records that his colleagues at the institute presented to him, together with an electrical record player. The quality of broadcast music in the early days of radio was not much better than that on the old records. In any case, did the Einsteins listen much to the radio?

HW: They listened quite a lot, but Herr Professor was less interested. Margot and I, we liked to sit in front of the box. In those days there were a lot of quiz programs that we took part in. Margot flattered me sometimes, saying: "Hertachen [little Herta], how do you do that? How did you come up with the answer?"

FH: What did they call you? Hertachen?

HW: That's what the daughters always called me. That can also be seen in Margot's letters after the war.

FH: And how did Frau Elsa address you?

HW: Herta, of course, as was customary. At first, Herr Professor did the same, but after about six months he called me Fräulein Herta. He intended this as a mark of respect. I recall it very clearly: On a certain occasion, everybody was looking for a particular book in his study, the tower room, and no one could find it. I went upstairs and found it right away. That was hardly a miracle because I always dusted all the books and had noticed where that book was located. Herr Professor said that I was better informed about his books than a secretary. Afterward, he began calling me Fräulein Herta, as a mark of respect, as he put it, and he stuck to this form of address. I recall very distinctly one time in Caputh, when he was sitting on the terrace and saw my mother walking up to the house, he called to me in the kitchen: "Fräulein Herta, mother is coming!"

FH: And I suppose you got along well with Frau Elsa?

HW: Oh yes.

FH: Was she not picayune, very petty?

HW: Not at all. But she was—how to put it—very exact. She was very precise with money, but not toward me. It was quite alright, for example, for me to order a half pound of coffee beans for myself every week, at her expense. If that was not sufficient, I had to pay for the rest myself. I have always been fond of real coffee.

FH: Did the Einsteins also drink real coffee, made from beans?

HW: After his heart ailment in the spring of 1928, Herr Professor drank only caffeine-free coffee, "Kaffee Haag." Apart from that, there was only tea, black tea. For visitors, as well. I would almost say that on such occasions, black tea was always served. Only Margot stuck to chamomile tea, on account of her gall bladder.

FH: I must ask you again: Didn't you have occasional clashes with Frau Elsa? You seemed to avoid the question before.

HW: Well, something like that is unavoidable, and it did happen occasionally. But whenever there was a discord, the two daughters did all they could to put things right again. Frau Dr. Kayser would come and bring me all kinds of things to make sure that I would not give notice. The only differences of opinion that occurred were with Frau Professor. Herr Professor paid no attention whatever to household affairs, and when there was a quarrel, he always took my side.

FH: When you came to Einstein, he had already grown a little stout, as opposed to his younger years, not to mention his student days in Zurich when he was often literally hungry because the paternal stipend was insufficient. Those years were long since in the past. In 1928, he wrote this comment under Emil Orlik's lithograph of him sitting in a chair, looking a bit plump and playing the violin:

> Thanks to science I have flourished,
> No fiddle player is this well-nourished.
>
> *Die Wissenschaft ist auch was wert*
> *Kein Geiger ist so wohlgenährt.*

HW: Yes, Herr Professor was quite well-nourished, one could say. In any case, I did not know him any other way. Frau Professor was also somewhat corpulent, but all the same she liked to eat pralines from among the presents she received. But she always gave some of them to me.

FH: One question regarding your employment status. You called yourself a housekeeper?

HW: *Stütze der Hausfrau* is what one called a housekeeper. I was the housekeeper.

FH: What kind of domestic workers were there at that time?

HW: There were cooks, chamber maids, nursery maids, general maids, and, yes, housekeepers.

FH: Was their pay different?

HW: As a housekeeper, your pay was better, and you were not obliged to do heavy work. Somebody else did that. A general maid, on the other hand, had to do all the work in the household.

FH: What was your salary?

HW: At first, in the Haberlandstrasse, forty-five Marks per month, then very soon after that, sixty Marks. In Caputh I got an extra ten Marks, because I had travel expenses and, also, for moving there with them.

FH: Was that a good salary for the prevailing circumstances?

HW: Yes, indeed. Compared to my friends, I was very well paid. You could get quite a lot for sixty Marks. A ready-made dress cost a little over twenty Marks, and, besides, my room and board were free. Moreover, in Berlin I often received tips from visitors and guests, particularly when they left late at night, after the doorman had gone to sleep. I would then take them down in the elevator and unlock the front door for them. They usually gave me something already in the elevator, or at the door, fifty Pfennigs or one Mark. That was customary. One Mark was a lot of money then. You paid sixty-five Pfennigs for a quarter of a pound of coffee beans.

FH: Einstein's guests were often very prominent people. I have Charlie Chaplin in mind, who was mentioned before. This super-wealthy film clown, who described Einstein's seven room apartment as "small and modest," must have given you a fat tip.

HW: I cannot recall today whether he gave me a lot. But I do remember another American, Goldmann or Goldemann; I cannot recall the name precisely. He was strictly kosher. He did not eat anything during his visits and drank only tea. The reason that I remember him so well is that every time he came he pressed ten Marks in my hand. That may not be much in dollars, but for me it was a lot of money.

FH: In his book *My Life*, Max Born mentioned a German-American, Henry Goldman, who owned a large private bank in New York.[8] In the 1920s he supported Born's institute with considerable funds, and he played the role of financial benefactor of many other German scientists, as well. Born introduced him to Einstein. He describes Goldman as a "charming old gentleman of rather Jewish appearance" who is unfortunately losing his eyesight.[9]

HW: Yes, that was him. He was seriously handicapped visually, or completely blind, and he always came to the Haberlandstrasse accompanied by a lady, perhaps his wife.

FH: How much free time did the housekeeper have?

HW: Once a week, I had the afternoon off, and every other week the entire Sunday. Often, when the Einsteins were away, I was off the whole Sunday and I was given two Marks to buy myself something to cook or go out to eat in a restaurant. Also, when the Einsteins were traveling for a longer time, I received two Marks per day for food, in addition to my salary. Quite often, I had the whole Sunday free and sometimes the whole weekend, when the family was invited somewhere and stayed overnight. I did not have to stay home all the time then.

FH: Did you receive a Christmas bonus or Christmas presents?

HW: Yes, the Einsteins were very generous in that regard. They always gave me bountiful Christmas presents. I received an extra fifty Marks, or material for two dresses, or linen for my dowry. They even set up and decorated a little Christmas tree especially for me. It stood on the sideboard in the dining room. And on Christmas Eve I had to eat the evening meal together with them, instead of alone in the kitchen. They wanted me to have a Christmas celebration, but I must admit that it was actually not so convenient for me because I was sometimes in a rush to get away. I usually went to the home of my cousin, where there were three children and always a very festive Christmas.

FH: I read in a biography that the domestic servant ate the evening meal together with the family every day and that Einstein served her. He was supposed to have said that someone who works manually all day is entitled to be served in the evening. Is that right?

HW: That is not correct. I always ate alone in the kitchen, as was customary, except on Christmas Eve.

FH: So this is one of those touching legends about Einstein that one comes across, particularly in American Einstein biographies. Back to Christmas Eve. The Einsteins did not celebrate Christmas, I suppose?

HW: No. But I did once go to the Hedwig Church with Margot, to a Catholic mass, to midnight mass. For the sake of the lovely music, not the churchy stuff.

FH: I suppose the Einsteins did not celebrate the traditional Jewish holidays either—for example, the Feast of Tabernacles [Sukkot].

HW: No, not at all.

FH: Although Einstein acknowledged being Jewish, he did not take these things seriously. Philipp Frank, whom I mentioned before, reported that in 1921, when Einstein was on a visit to Prague, an orthodox Jew asked him where there was a kosher restaurant. Einstein named such a restaurant, but when the practicing Jew wanted to know if it was really strictly kosher, Einstein told him vehemently, "Only an ox eats really strictly kosher. For he eats grass." He could only scoff at the dietary ritual of the Jews, and even in his parents' home no attention was paid to these customs.

HW: I can hardly say that the household was kosher; they ate everything.

FH: I mentioned before how variably different visitors judged the Haberlandstrasse apartment; coming back to it, did the Einsteins have upholstered chairs, or mostly ordinary chairs?

HW: The chairs around the large dining table were all plain chairs. Elsewhere there were armchairs and plain chairs, and, in the living room, a sofa and upholstered chairs with rounded back rests (i.e., in the Biedermeier style). Apart from that I can recall no upholstered furniture, except for the small sofa and a plain chair in the eating niche.

FH: Did Einstein use an office chair at his desk?

HW: When Herr Professor was writing he always sat in an armchair with a high back. Upstairs, in the tower room, there was also a high-backed armchair at the desk. Downstairs, the Einsteins sat in the high-backed armchair generally only for reading. In the library, there was a plain chair at the desk.

FH: Do you recall any clocks in the apartment? Was there perhaps a grandfather clock that melodiously struck the hours, or a long pendulum clock in a glass case, the kind that belonged in a proper bourgeois home decor—the kind that is very desirable again today and very expensive? A "regulator clock" would have fitted in well with the heavy furniture.

HW: I cannot recall such a clock. In the Biedermeier room, a clock sat on a high ledge that looked a little like a mantel piece. It seemed to me to be a gilded clock, but it might just have been bronze.

FH: When did you have to get up every day?

HW: Around seven o'clock. Not much earlier. I then vacuumed the living and dining rooms lightly. Breakfast was not too early, certainly not before eight or half-past. I did not have much to do at breakfast. I had to make fried eggs and scrambled eggs almost every day, also for Herr Professor. Margot always ate stirred eggs, very soft and fluffy.

FH: Did you use fireproof cooking utensils, like today's Jena glass?

FH: We did not have such glass utensils. We did have fireproof ceramic forms that resembled porcelain. Only the coffee machine was glass, and Herr Professor prepared his caffeine-free Kaffee Haag every morning in that machine. It had a little alcohol burner, and once it was closed up the coffee rose into the two glass globes. How that works, I can no longer recall, but in my mind's eye I can still see that little flame being lit.[10]

FH: You mentioned earlier that you vacuumed the dining and living rooms. I take it that vacuum cleaners already existed then.

HW: Yes, they were already around in the twenties. They were not as convenient as today, a large bag was mounted on the outside, but that's what the first models were like.

FH: So the day began with breakfast at eight or half past. How did it go afterward? Did Einstein travel regularly into the city, to the Academy or the University?

HW: Not every day. With him everything was quite irregular. One day, Herr Professor would travel to an institute, another day he worked in the tower room; it was very variable.

FH: So no one can say that Einstein left the house "by the clock" and returned at a particular time, as is said of other scholars, as for instance of the great philosopher Immanuel Kant in Königsberg. Kant took a daily walk with such regularity that the citizens set their clocks according to him. It was known to the minute when Professor Kant passed a particular street. At a time without radio, such a live "time check" must have been quite useful. But Einstein would have been unsuitable for that.

HW: No, everything was quite irregular. Sometimes Herr Professor went up to his study in the morning and worked there for many hours, sometimes he stayed in his bedroom all morning. He did come out to eat breakfast, and then he went back to cogitate about something. Sometimes he sat in the library, and sometimes he played his violin or the piano for several hours. Everything was quite spontaneous; there was no such thing as a regular daily routine for him.

FH: An English Einstein biography says that Einstein went to his office in the Academy of Sciences every day. Archival documents show, however, that even though he was full member of the Academy, Einstein did not have a room in the Academy building. He did, however, attend the Academy sessions regularly, and he often gave lectures and talks at the University of Berlin. He also participated in the colloquium that, during the term, met every Wednesday at the physics institute on the Reichstagufer [Street]. These colloquia brought together physicists, geophysicists, and astronomers for an exchange of scientific ideas, and to attend them he had to travel to the city center and had to be on time, of course.

HW: Herr Professor did travel into town on some days, just not every day.

FH: How did Einstein dress?

HW: Very plainly—plainly and modestly. I must have told you already that I had to cut the cuffs of his shirt sleeves so that they would not always show. Cuff-less shirts did not exist in those days; they all had such long sleeves that it bothered him. I had to shorten the sleeves and sew them up. He wore cuffs only on special occasions. I remember an occasion when Herr Professor was supposed to get a new suit but could not be induced to go to the tailor or to the clothing store. That's when Frau Professor took an old jacket that fit him out of his closet, brought it to a large clothing store on the Gertraudenstrasse, and bought a suit off the rack. She took the jacket to get the right size, because Herr Professor could not be persuaded to come along. He kept saying that it was unnecessary, the old suit will still do, and I don't need a new one.

FH: The Berlin journalist and Einstein biographer Antonina Vallentin is therefore right when she writes in her book *The Drama of Albert Einstein*: "He greatly prefers wearing something that is familiar from long usage, over a fabric whose touch he is not used to. He is more comfortable in an old dressing gown full of holes, than in a new one that was given to him." The author of this biography, whom we will meet again in our talks, may be known to you as Frau Luchaire. Her husband was a diplomat who worked for the League of Nations in Geneva; he chaired a commission Einstein belonged to.

HW: I remember the name Luchaire.

FH: What did Einstein wear at home? Photos taken in his later years often show him wearing a sweater.

HW: That is the sort of thing he liked to wear, particularly later on, in Caputh, but also in the city apartment.

FH: What if friends were expected—for example, colleagues like Max Planck or Max von Laue?[11]

HW: Then he would usually wear a shirt with a collar and a jacket or a cardigan. That was often the case.

FH: He was apparently reluctant to put on a tie.

HW: Yes, but he did own ties, and he wore them occasionally. Otherwise, he preferred wearing a cardigan and no tie. Also, when he walked up the stairs to work, he usually wore a cardigan. He always had to use the stairs; the elevator did not go up to the tower room.

FH: What was the Einsteins' social life like? Einstein referred to himself on several occasions as a "typical loner" (*Einspänner*) who felt little need for people or human society. Did social events appeal to him at all?

HW: I recently heard on a radio program that Professor Einstein liked attending gatherings and that he enjoyed being seen in public. Well, all I can say is that, most of the time, the opposite was the case. Often, when

Frau Professor had arranged something or other, he went along but only reluctantly. I never witnessed him showing off or enjoying speech making. I only know that, now and then, Herr Professor played the violin at a benefit event. Of that I am certain. But that he liked being the center of attention? That is simply incorrect. There were in fact frequent quarrels between him and Frau Professor when she had accepted yet another invitation on his behalf. I know for certain that Herr Professor disliked attending these events and that he often railed very angrily against her for having accepted the invitation.

FH: But Frau Elsa apparently placed great store on her famous husband being often seen in public.

HW: That she did, but as for herself, she did not always accompany him.

FH: Talking of social functions, did Einstein own a tail coat? Or did he borrow this garment for particularly festive occasions from an acquaintance of approximately the same size?

HW: He certainly had a frock coat, which was customarily worn on social affairs when he lived in Berlin. That I know. But I cannot recall a tail coat. As for the suggestion that he might have borrowed one—well, that I consider improbable.

FH: Antonina Vallentin writes in her biography that Einstein once said: "Tail coat? But why a tail coat? I never owned one and managed very well without one."

HW: That is quite likely.

FH: But he must have owned a tuxedo. He can be seen wearing one at festivities—for instance, when he gave his Nobel address in the presence of the Swedish king, in Goteborg, in the summer of 1923. It was reported at the time that Einstein left an unfavorable impression because he wore a rather shabby tuxedo.

HW: Herr Professor paid very little attention to ceremonial clothing. Frau Professor, on the other hand, always wanted to have beautiful clothes, but she could not wear some of them because she no longer had the necessary figure. To save money, she had many of her clothes made by a seamstress who came to the house regularly. She would stay for several days and worked in the guest room, but what she made was not really as chic as what was worn by the ladies who came to visit. The seamstress just couldn't do it that well.

FH: Did Frau Elsa never go to one of the great fashion houses to shop for her wardrobe?

HW: That, never! It was all left to the house seamstress, who was much cheaper. In the late summer, the seamstress sometimes also came out to Caputh, where she worked in Margot's room.

FH: Did Frau Elsa have a well-stocked clothes closet?

HW: Actually, Frau Professor had quite a small wardrobe. But when the two returned from America in the spring of 1931, she, as well as Herr Professor, wore many things that I had never seen before. I am pretty certain that these were presents, at least, in part.

FH: That may well be, but the California Technical University in Pasadena, called Caltech, certainly paid Einstein an honorarium for his teaching activities, and that would pay for quite a few things.

But now for something entirely different. Several biographies report that Einstein always had only very little pocket money on him. Is that an exaggeration?

HW: That is not exaggerated; that is correct. I once overheard a heated argument between Herr Professor and Frau Professor. He said that since his theater ticket had been a present and a car was picking him up, he wanted to have at least enough money to pay the coat-check charge for Toni and himself. The quarrel was so loud and heated that I could not help overhearing it. It took place in the dining room or in the hall, because Herr Professor was already on his way out.

FH: You mentioned a certain Toni just now. Who was she?

HW: Frau Toni Mendel, always called Toni by everyone, was a very good-looking, attractive woman, a fine lady, I would like to say. She was perhaps a little younger than Frau Professor. She was very wealthy, and when she came she always brought pralines and other things for Frau Professor. She had a large, beautiful house on the Wannsee, and I was often there when the Einsteins were visiting. Even before she met me in person, she sent me a pot of flowers from her greenhouse because she supposedly had liked my voice over the telephone. Her chauffeur delivered it to me when he came to pick up Herr Professor. She and Herr Professor had a close friendship, which Frau Professor respected, in my opinion, only because she was obliged to. It was a friendship of many years with the whole [Mendel] family. Unfortunately, I know nothing more of Frau Mendel's personal circumstances. I believe she was already widowed. But as to the pocket money, that is correct, I can confirm it.

FH: That is really astonishing because Einstein had a considerable income. As a full member of the Prussian Academy of Sciences he earned a high salary. He was, furthermore, compensated as the director of the Institute for Physics of the Kaiser Wilhelm Gesellschaft, although that institute existed only on paper at that time; by the time it was built, Einstein had long since been in Princeton. How do you explain the niggardliness of Frau Elsa?

HW: I suspect that the apartment cost an awful lot. There were seven rooms, after all, and in addition the smaller rooms and the study in the attic. Apartments in the Bavarian quarter were very expensive, as far as I know. But partly it was a little in Frau Professor's character.

FH: The biographies that maintain that, in his Berlin years, Einstein had to account for every Pfennig he spent—are they right then?

HW: Yes, Herr Professor had very little money at his disposal.

FH: The scene Antonina Vallentin describes in her book—she witnessed it herself—may therefore be relevant. She reports that on a visit in Caputh, she saw Einstein going through the pockets of his old white trousers, in search of a certain piece of paper. In doing so, he unfolded a piece of paper with a poem that Queen Elisabeth of Belgium had dedicated to him. She noticed that in the corner of this ivory-colored sheet there were a few words and numbers in Einstein's small, regular handwriting. Looking at them more closely, she was astonished to read: bus, 50 Pfennigs, newspaper . . . paper . . . So, her account might not have been made up, although certain things in Frau Luchaire's book are demonstrably untrue.

HW: This report is probably correct. Herr Professor was always short of cash. In any case, the Einsteins lived very modestly and simply. I cannot recall Herr Professor ever drinking alcohol. It played no big role in the household. To be sure, Professor Plesch sometimes sent sumptuous gift baskets that contained a bottle of good cognac; Frau Professor took the bottle immediately to the pantry and locked it up in a special cabinet, for which only she had the key. Supposedly, there was a fifty year-old cognac in the cabinet, as well as some very fine cigars. These were only passed around on very special occasions. At such events a little cognac was drunk, as well. In those days, it was in any case not as common to have a drink at every opportunity, as it is today.

FH: What you say is consistent with Einstein's reply to an American reporter who asked what he thought of Prohibition—it prohibited the production, importation, and consumption of alcoholic beverages—which was still in force in the United States in 1930. It is reported that Einstein answered with a laugh, "I don't drink, so it makes no difference to me."

HW: Yes, that is right. I cannot recall ever seeing champagne at the Einsteins, for example. I did not find out about champagne until later.

FH: In his memoir, Plesch placed limits on Einstein's position on alcohol. He thought that Einstein was neither a drinker, nor a fanatical teetotaler and that he occasionally liked to nip a glass of cognac, but no more than that. Would you agree?

HW: Yes, but he nipped at a glass of cognac very rarely.

FH: Was the ambience you found at the Einsteins different and better than before, at the home of Geheimrat H.?

HW: It was decidedly different and much, much better. I was also accorded far more respect as a person. In the previous job, I was a real domestic servant, and I was exploited a great deal. It was quite different at the Einsteins. For the first anniversary of my moving into the Haberlandstrasse, I was allowed to invite guests. That also had something to do with the Einsteins having had so little luck with their domestics before and, also, because they were all very fond of me. I was permitted to invite visitors; my brother and sister-in-law came from Berlin and my cousin and her husband from Schöneweide. Frau Professor donated a bottle of wine and a cake. Herr Professor went up to his study and Frau Professor withdrew to the living room, so that we could celebrate in the dining room.

FH: Was this jubilee celebration repeated every year?

HW: No, but I was allowed to invite guests at any time, particularly later, out there in Caputh, where we made ourselves comfortable upstairs on the sun terrace. As far as compassionate kindness is concerned, the Einsteins were tremendous. That I would like to emphasize explicitly because, previously, I had had the opposite experience. One time, when I had to spend several weeks in the hospital, they gave me for Christmas not fifty Marks, but a hundred, and after I was released they let me stay with my parents in Lautawerk to recover while they were in America. My salary continued to be paid in full. Afterward, the Einsteins hired a cleaning lady every other day, so I could take it a little easier. They were always kind to the members of my family who came to visit me. When my younger brother came for the first time, and they did not know what to give him, they gave him the busts of Goethe and Schiller that he had admired. Afterward, my brother held the busts in great esteem and told everyone that they came from Professor Einstein's apartment.

FH: We have already mentioned how nearsighted Frau Professor was and that her vanity kept her from wearing spectacles.

HW: Frau Professor often mislaid her keys, and then I had to look for them. She would put them down somewhere, and then she could not find them. Her nearsightedness troubled her a lot, but under no circumstances would she wear glasses.

FH: Frau Vallentin has made some interesting observations in that connection. She writes that Elsa had a look of "utter helplessness" on account of her extreme nearsightedness, that it created "a world of terror" around her. With her head down, she stepped into the path of people that she failed to recognize, she bumped into objects, and she used her lorgnette to examine what was on her plate. At a banquet in the United States, Frau Elsa, supposedly, started to cut up an orchid on her plate because she mistook the table decoration for the salad. Although this might well be an exaggeration, or even a fabrication, this anecdote illustrates tellingly the deplorable situation Frau Einstein found herself in.

HW: It made Frau Professor intensely unhappy.

FH: Since some Einstein biographies portray Albert and Elsa Einstein's life together as an unblemished family idyll, I would like to ask you: Apart from the quarrels about pocket money, weren't there other serious disputes, as well?

HW: Yes, that did happen, and it was almost always about women. Herr Professor just liked to look at beautiful women; he always had a weakness for lovely ladies. I'll be able to say more about that when we talk about the time in Caputh. But on the other hand, beautiful women also liked to be seen with Herr Professor, and Toni Mendel, whom I mentioned earlier, was among them.

FH: In your time, was Einstein still in touch with his first wife, Mileva, from whom he had been divorced in Switzerland in the beginning of 1919? In his letters to friends, he referred to her as the "lapsed one" (*die Verflossene*).

HW: She came to Berlin a few times, but, as I recall, alone, without the two sons. She did not stay in the Haberlandstrasse but only came to visit. Herr Professor sat with her in the library, and they had a friendly conversation for an hour or two. Frau Professor also joined in.

FH: Did you have any dealings with Frau Mileva?

HW: No, I just opened the door for her, probably also said goodbye to her or brought her downstairs in the elevator. I had no other contacts with her.

FH: What was your impression of her?

HW: She was quite tall and quite slim, but she was no beauty. Frau Professor was decidedly better looking. Frau Mileva was a different type, a somewhat exotic type.

FH: She came from Serbia, from the Banat region, from a family of farmers. That may have to do with it.

HW: Aha, I never knew that.

FH: Did the two sons, Hans Albert and Eduard, come for only short visits, or did they sometimes stay with their father for longer periods?

HW: That was variable. Sometimes they came for a longer time.

FH: Both of them together?

HW: As far as I can recall, they did not come at the same time, but they did stay in the Haberlandstrasse, in the guest room. I remember several visits of Eduard very well; I recall the terrible music he made. The way he furiously pounded the keys and abused the piano, it was horrendous. Completely crazy. He struck me as being overly excited. The elder son, who was always called Albert, not Hans Albert, continued to visit when he was already married, but then only in Caputh. Contrary to his nervous brother, he seemed calm and solid.

FH: Hans Albert was born in 1904 in Bern and studied engineering science in Zurich, where he married a Swiss woman while still quite young, around 1929. In the Fall of 1930, Einstein commented in a letter—he was then fifty two—that his oldest son had "disrespectfully promoted him to grandfather." Albert worked in Germany for a few years before he, too, emigrated to the United States. When last heard of, he had an appointment as professor of hydraulics at the University of California, in Berkeley. He died there in the fall of 1973. Einstein's grandson Bernhard is also an engineering scientist.

HW: That I did not know. But I did know that Herr Professor had a grandson, and I remember Hans Albert and his wife very clearly; they impressed me as being uncomplicated and modest. They stayed in the guest room in Caputh. It was probably in the summer or fall of 1932. Whether little Bernhard was with them, I can no longer say.

FH: He must have been in Caputh because a photo, apparently taken on the lower terrace of the summer house, shows Einstein with Hans Albert and a flaxen-haired boy of about two years. But that is already a preview of our next conversation, in which we will chat about Einstein's visitors and guests. But for today, one more question regarding the city apartment: Can you recall any pictures that decorated the walls of the living room or the other rooms?

HW: In the dining room there was a large painting; it may have been a valuable original painting, but I can no longer say what it showed. In the past, I did not take much interest in pictures, but when I now see old pictures, it sometimes come to mind. It was framed without glass and was somewhat darkened with age, a genuine oil painting.

FH: One Einstein biography describes a large picture in the city apartment in considerable detail, but I prefer not to replicate that description here. Such particulars are often inexact or invented from A to Z. We have no wish to help spread legends about Einstein, but rather, the truth.

SECOND CONVERSATION:
FREQUENT VISITORS AND RARE GUESTS

FH: I would like to add something to our previous conversation. You mentioned a small telescope on a tripod that stood in Einstein's study, next to his desk. Since Einstein had no particular connection to observational astronomy, it is unlikely that he acquired this amateur's telescope to make celestial observations. If he wanted to make such observations, he could easily do so with the large telescopes of the Potsdam observatory. It is more likely that the little telescope had once belonged to his uncle and father-in-law, or that it was a present from a friend. That brings Friedrich Simon Archenhold to mind, the founder and director of the Peoples' Observatory in Berlin-Treptow.[12] Shortly before the First World War, Archenhold had designed a handy school telescope, which was manufactured commercially by a firm in Munich, and it is not impossible that he presented Einstein with one of their telescopes. In 1915 Einstein gave a popular science lecture on the theory of relativity in the large lecture hall of the Treptow Observatory. It was the first such lecture since he moved to Berlin, and a friendly present of this kind may well have been justified, but it is impossible to know if it happened. What seems certain, however, is that Einstein did not buy it for himself.

Let us now leave the realm of astronomy and delve into that of gastronomy. Did you do all the cooking at the Einsteins on your own, or did Frau Elsa sometimes join you at the stove? She might have been fond of particular specialties from her Swabian homeland that you, as a Berliner, would not know.

HW: No. What Frau Elsa always did take on was the peeling of asparagus. She told me that in the boarding school where she was educated she had always been the best asparagus peeler. That is why she always peeled the asparagus herself, to make sure that it was quite tender.

FH: In view of her nearsightedness—consider the anecdote of her cutting up orchids on her plate—that must have been quite a trick. She must have worked by touch.

HW: We used to eat a great deal of asparagus, particularly in Caputh, for it was in an authentic asparagus-growing region. We used to buy asparagus very often, and she would always say right away, "I'll take care of the peeling."

FH: You told me that Einstein enjoyed his food.

HW: Yes, that is true. He ate with zest and pleasure [*mit Lust und Liebe*], I would say. It did not have to be anything special.

FH: Was it customary to eat a second breakfast, a so-called *Gabelfrühstück*, between the morning coffee and lunch?

HW: No. There were eggs for breakfast already, usually fried. There was also a lot of honey, which we bought by the pail. In September, when the blossoms on the heath were finished, a beekeeper used to come and bring his dark honey. There was always honey at breakfast.

FH: Did the Einsteins eat dark bread or white bread, or rolls—*Schrippen* as they are called in Berlin?

HW: Rolls. They were delivered to the house gratis, and every morning they hung on the back door in a paper bag. Milk was also already in front of the door. In those days, everything was still delivered to the house, free of charge.

FH: What were the customary beverages at the Einsteins? Alcohol was virtually never consumed. Was there bottled water at mealtimes?

HW: No. Only if there was a larger circle of friends was wine consumed or a punch bowl prepared. It was almost always a bowl of celery punch.

FH: I am familiar with strawberry punch, but I have never heard of celery punch. However, I am not knowledgeable in these matters.

HW: Yes, it was a celery punch. I also heard of it for the first time at the Einsteins, but its taste was delicious, very savory. Frau Professor always assisted in its preparation.

FH: Was fruit popular?

HW: We bought a lot of fruit. Many apples and pears and, of course, strawberries.

FH: What was served at afternoon tea? Cake, cookies, patties?

HW: In general, nothing was eaten with the tea. Neither was there any cake, which we fetched only when guests came. We used to buy it at the pastry shop on the Bayrischer Platz. Occasionally, we baked for afternoon tea, an apple pie like I had never seen before, with a lattice made from sour cream and egg and without a top, like we do it here. That must have been a Swabian way. Occasionally, also a cherry pie, also with that kind of lattice on top. In the beginning, Frau Professor helped a lot, so I would get to know her preferences, but later I did everything by myself.

FH: Did the daughters take part in the household work? At least Margot, who resided there during your time?

HW: No. Not at all.

FH: You told me once that the Einsteins ate a lot of vegetables and used few spices.

HW: Yes, particularly after Herr Professor's illness in the spring of 1928. The attending physician, Professor Plesch, ordered a salt-free diet, and we stuck to it strictly.

FH: When was the evening meal eaten?

HW: Sometime between six and seven, definitely not very late. Usually there were cold cuts, cheese, and eggs.

FH: The Einsteins, apparently, liked eggs and ate a lot of them.

HW: There were eggs already at breakfast, as I said before. Herr Professor always ate two fried eggs, at least two.

FH: You told me that the Einsteins had a regular egg delivery man.

HW: Yes, a Jewish man came and brought fresh eggs. Dear me, he was so grubby! Because of his age, he did not have to use the delivery staircase in the back but was brought up in the elevator by the porter. Frau Professor had arranged that. The Einsteins bought eggs from him to support him, but how they knew him I cannot say. In Berlin, he came regularly, but he did not come out to Caputh. There we bought the eggs in the village.

FH: We seem to have reconstructed the menu of the Einstein household to some extent, though we may have overlooked this or that. That doesn't matter; we don't strive for completeness in these minor matters. Since some Einstein biographies cover this subject quite extensively, however, I think it's appropriate to inquire about some culinary particulars. An American biography says, for instance, that during his Berlin period Einstein was very fond of eating soup and sausages. When Frau Elsa

Einstein as high school student, with his sister, Maja. Munich, ca. 1893. *Courtesy of LBI.*

Einstein with friends in 1899, the year he graduated from ETH. L to R: Marcel Grossmann, Einstein, Gustav Geissler, Eugen Grossmann (Marcel's brother). *Courtesy of Hebrew University of Jerusalem, Albert Einstein Archives; AIP Emilio Segre Visua Archives.*

Einstein's sons, Eduard (7) and Hans Albert (13), hiking in Arosa, Switzerland, July 1917. *Courtesy of the Leo Baeck Institute.*

Einstein with Berlin colleagues, Dahlem, 1921. L to R: Hertha Spooner, Einstein, Hugo Grotrian, Ingrid Franck (wife of James Franck), Wilhelm Westphal, James Franck, Otto von Bayer, Lise Meitner, Peter Pringsheim, Fritz Haber, Gustav Hertz, Otto Hahn. *Courtesy of AIP Emilio Segre Visual Archives, Aristid V. Grosse Collection.*

Mileva and Technical Expert Class 3, ca. 1905.
Courtesy of HIP / Art Resource, NY.

Elsa and Einstein on their visit to Washington, where they were feted by President Harding, 1921. *Courtesy of Library of Congress, Prints & Photographs Division, photograph by Harris & Ewing.*

Einstein playing chamber music. Lithograph by Emil Orlik, 1928. *Proofsheet, dedicated to Mr. and Mrs. Schwartz. Courtesy of AIP Emilio Segre Visual Archives.*

The futuristic astrophysical laboratory, known as Einsteinturm, on the Telegraphenberg, Potsdam, 1928. *Courtesy of bpk, Berlin / Art Resource, NY.*

Piano trio aboard the SS *Deutschland*, somewhere between New York and Hamburg, March 1931. *Courtesy of bpk / Art Resource, NY.*

Einstein and fellow physicists deliberating in Ehrenfest's home, Leyden. L to R: Einstein, Paul Ehrenfest, Paul Langevin, Heike Kamerlingh-Onnes, and Pierre Weiss. *Courtesy of HIP / Art Resource, NY.*

Einstein in his tower study (*Turmzimmer*), 1927. On the wall is a picture of Newton. *Courtesy of bpk / Art Resource, NY.*

Einstein in the library of the Haberlandstrasse apartment, 1927. Note the telescope to the right of the desk. Einstein's usual study was the tower room, constructed at his behest in the attic in 1922. *Courtesy of the Leo Baeck Institute.*

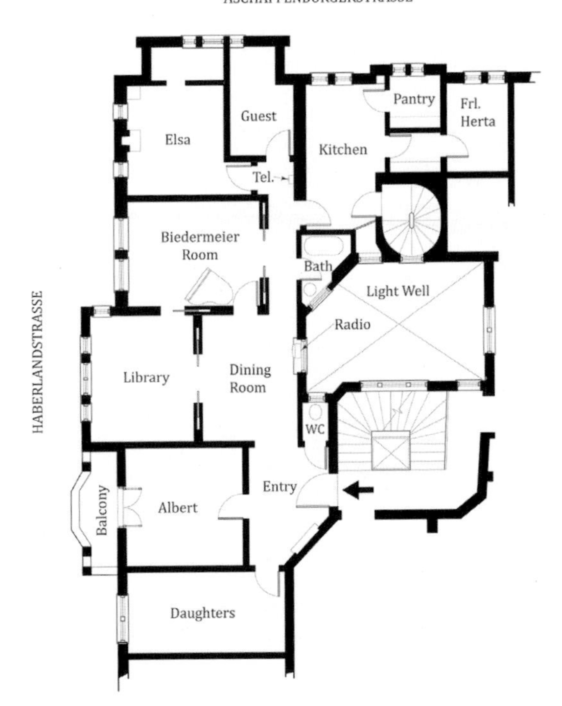

Probable layout of the Einstein apartment at Haberlandstrasse 5. The floor plan was reconstructed by Simon Eisinger, AIA, based on original architectural drawings of the building, interior photographs, and descriptions by eyewitnesses.

A 1909 postcard showing the apartment building at Haberlandstrasse 5. The location of the Einstein apartment has been highlighted. *Photographer unknown. Public domain.*

Pianist Joseph Schwarz, violinist Boris Schwartz (his son), and Einstein during one of their periodic trio sessions in the Biedermeier room. *Courtesy of the Leo Baeck Institute.*

View from the Biedermeier room into the library. *Courtesy of the Leo Baeck Institute.*

Einstein in Elsa's bedroom in the Haberlandstrasse apartment, 1932. *Photo © Hulton-Deutsch Collection / CORBIS.*

János Plesch and Einstein in Plesch's Gatow estate, 1929. *Photo © Bettmann / CORBIS.*

Einstein and his elder son, Hans Albert, in the library of the Haberlandstrasse apartment, 1927. *Courtesy of Granger, NYC.*

Margot Einstein and Dmitri Marianoff on their way to their wedding, accompanied by Einstein, Berlin. December 1930. *Courtesy of Scala / Art Resource, NY.*

Einstein's summer home in Caputh, ca. 1930. The three French doors of the living room can be seen underneath the upper terrace. *Courtesy of the Leo Baeck Institute.*

Einstein and Adolf Harms, the builder of *Tümmler*. Harms had just delivered Einstein's grand birthday present to him, its bow decorated with a wreath. Caputh, August 1928. *Courtesy of Granger, NYC.*

Einstein and his friend Hermann Anschütz-Kämpfe, sailing on Kiel Bay. *Courtesy of the Leo Baeck Institute.*

Rabindranath Tagore and Einstein descending the steps from the summer home in Caputh. Behind Einstein, Margot. *Courtesy of the Leo Baeck Institute.*

Herta Schiefelbein with Purzel, on the upper terrace of the Caputh house, ca. 1930. *Courtesy of Friedrich Herneck Archive, Dresden.*

Friedrich Herneck on his seventy-fifth birthday, 1984. *Courtesy, private archive of Dieter B. Hermann.*

Niels Bohr and Einstein in conversation at Ehrenhaft's home. Leyden, ca. 1930. *Photograph by Paul Ehrenfest. Courtesy of AIP Emilio Segre Visual Archives.*

During his stay in Pasadena, Einstein visited the Mount Wilson Observatory, where he was shown the spectroscopic evidence that our universe is expanding. In front, L to R: Einstein, Edwin Hubble, Robert Millikan. *Courtesy of Scala / Art Resource, NY.*

Fräulein Herta Schiefelbein
war vom 1. Mai 1927 bis 1. Juni
1933 als Stütze in meinem
Hause.
Sie ist durchaus ehrlich, zuverlässig,
geschickt in allen häuslichen Arbeiten.
Im Kochen sehr bewandert.
Ihre Entlassung erfolgt wegen Übersiedelung
ins Ausland. Meine besten Wünsche
begleiten sie.

Frau Albert Einstein

Coq sur mer Mai 33.

Elsa Einstein's reference for Herta Schiefelbein, dated May 1933. It reads: "Fräulein Herta Schiefelbein worked in my home as housekeeper from 1 May 1927 till 1 June 1933. She is absolutely honest, reliable, and skilled in all domestic arts, and a very capable cook. The severance is due to a move abroad. My best wishes accompany her. Frau Albert Einstein, Coq sur Mer, May, 33." *Courtesy of Friedrich Herneck Archive, Dresden.*

Liebe Herta!

Sie hätten die Freude sehen sollen, als Ihr Brief ankam. So oft fragten wir uns: Was ist denn aus unserer guten tapferen Herta geworden? Und wir fanden es bedenklich, dass wir so lange nichts hörten. Sie haben ja so viel und so Verschiedenartiges mit uns zusammen erlebt, in der nun verschwundenen Haberlandstrasse und in Caputh. Das Häuschen dort haben nun die Russen und ein Nazi hat das dicke Segelschiff bekommen. Hoffentlich geht's bei Ihnen bald wieder besser mit dem leidigen Futter.

Alle herzlichen Wünsche
Ihr
A. E.

Einstein's letter to Herta Waldow, 1947. *Courtesy of Friedrich Herneck Archive, Dresden.*

Einstein at sixty-three, in his Princeton study. *Photo by Roman Vishniac. Copyright Mara Vishniac Kohn, Courtesy of the International Center of Photography.*

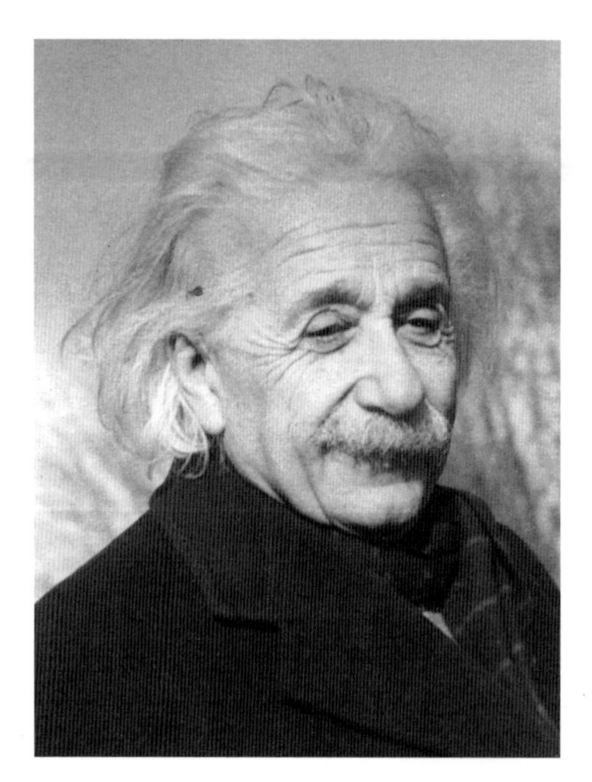

Einstein at seventy. *Photo by Alfred Wyler. Courtesy of Marianne D. New.*

pointed out to him that he had eaten the same dish the day before, he is supposed to have answered, "Doesn't matter." Another Einstein biography, also published in the United States, states the Einstein's favorite dish was "pork sausage," which presumably refers to Berlin's *Bockwurst*.

HW: I would have known about that. I cannot confirm that Herr Professor was fond of Bockwurst, but what he did like very much was beans, young green string beans.

FH: That is the dish with which you made a favorable first impression.

HW: He was also very fond of eggs with mushrooms, particularly porcini and other boletes that he often brought back from his walks in the woods. If it were up to him, he would probably have eaten mushrooms three times a day, that's how fond he was of them. He also liked salads, rice, and spaghetti prepared in the Italian manner. But in general, Herr Professor was undemanding in what he ate. Schnitzel or meatballs were rare. Filet or steak had to be well done, it could not be bloody; otherwise, he would not eat it. He would then always say, "I am not a tiger."

FH: Was Einstein in the habit of taking a midday, after-dinner nap, the kind that the physicist Hermann von Helmholtz strictly adhered to? He slept on a leather sofa, in his study in the Physics Institute of the University on the Reichstagufer. His successor guarded this sofa religiously, and it was left as it was until a new covering became unavoidable. Did Einstein also take a noontime nap, regularly?

HW: Not regularly. Herr Professor may occasionally have rested after the midday meal. He withdrew into his bedroom, but he may have done some reading or writing there; that I have no way of knowing.

FH: When did Einstein go to bed in the evening? Some intellectuals, scholars, and writers sit at their desk till late into the night, when they are most productive; they are jokingly known as night owls. Did Einstein belong to the night owls?

HW: If he was not invited somewhere, or attended a meeting, he withdrew to his bedroom right after the evening meal. He continued to work there, probably in bed, because in the morning there was a pile of notes on his bedside table that he had written his calculations on. But it was very variable.

FH: It has been reported that Einstein did not wear pajamas and this was confirmed by the Polish physicist Leopold Infeld, who collaborated with him in Princeton. Was this also the case in his Berlin days?

HW: In Berlin, Herr Professor had pajamas and nightshirts and he must also have worn them, because they were washed again and again. This was not done by me, but by the woman who came to do the laundry. A lot of the laundry was also done by a large commercial laundry that picked it up and delivered it, wardrobe-ready.

FH: Did the Einsteins often go to concerts or to the theater in the evening?

HW: Less to the theater, perhaps, but very often to concerts. Herr Professor loved going to concerts.

FH: With Frau Elsa or Margot, or with Toni?

HW: Until she was married, he often went with Margot, also often with Frau Mendel. Frau Professor went along only occasionally.

FH: What did he wear on these occasions?

HW: A dark suit. But getting dressed up was always problematical for him. Whenever he had to get dressed for a big occasion, there were always arguments. It was something he disliked passionately.

FH: So, Plesch is right when he writes about Einstein, "Changing into evening clothes is always agony for him. He does it anyway because as a conscientious person he accepts his obligations, including this one. But it is one reason that he declines most invitations to big events."[13]

HW: That I can confirm.

FH: Was Einstein usually picked up by a car to go to his functions, or did he use public transportation?

HW: Very often he was picked up by car. Chief Physician Professor Katzenstein often put his car and driver at his disposal. Toni Mendel also often sent her automobile.

FH: Did the Einsteins often go to the cinema? It says in one biography that Einstein once made a special trip from Caputh to Potsdam to see a particular film. When you joined the household some famous silent films were being shown, with performers like Charlie Chaplin, Emil Jannings, Paul Wegener, and others. Several famous Soviet films, by Eisenstein and by Pudovkin, were also playing, such as *Battleship Potemkin*, *Ten Days that*

Shook the World, *Mother*—based on Gorki's novel, *Storm over Asia*—to name just the most important ones. Sound films became widespread around 1930. Do you know if the Einsteins liked to watch films?

HW: That I cannot recall.

FH: Did they sometimes go to cabaret performances? For example, to Otto Reutter or Claire Waldoff, or perhaps to Caro's Laughter Stage [*Lachbühne*]? Plesch wrote that Einstein could "laugh convulsively at the most harmless matters."[14] That being so, the *Lachbühne* would have been the right place for him.

HW: I, personally, liked to go to Caro's *Lachbühne* on the Weinbergsweg. But not the Einsteins. They went to lots of concerts, also to the opera, and to dinner parties. Invitations virtually rained down on them. There were far more than they could accept.

FH: Now I have a few questions about their visitors. Did the porter telephone the apartment when a visitor came to see Einstein? How did this work?

HW: The porter brought the visitors up in the elevator and to the apartment door. I opened it, took their coats, and announced them.

FH: Where did you take the visitors? To the living room or the library?

HW: Usually to the living room, but if the visitors came only to see Herr Professor, then I took them to the library where they would talk.

FH: Only a few of the visitors went upstairs to the study, I suppose?

HW: Only very rarely did someone come upstairs. It would have had to be a close friend or a colleague; for example, Geheimrat Planck or Professor von Laue.

FH: In the transcript of your police interrogation on September 5, 1934, you stated that Planck was among the most frequent visitors of Einstein.

HW: Yes, Geheimrat Planck often came to the house.

FH: I was fortunate enough to have attended the lectures of that famous physicist and Nobelist, who started quantum theory at the turn of the century. He left a strong, unforgettable impression, even when he seemed somewhat formal and "academic," virtually like the "Nobelist incarnate," as a witty German-American physics colleague aptly characterized him. You doubtlessly still remember this unpretentious, modest scientist.[15]

HW: I would immediately recognize Geheimrat Planck again. He came often and there were also many photos of him in newspapers and magazines. I also saw many pictures of him after the war.

FH: Planck's former student, Max von Laue, also a Nobel Prize winner in physics, was also a frequent guest in the Haberlandstrasse and also visited Einstein several times in Caputh, as he himself told me.[16] Laue was among Einstein's closest friends in Germany. A former student at Berlin University described to me how, in the mid-1920s, he once saw Einstein and Laue walking arm-in-arm in the garden in front of the university building at Unter den Linden. Einstein cherished the discoverer of X-ray interference, not only as a brilliant mathematician and theoretical physicist, but also as a man of character with a highly developed sense of justice. During the Nazi period, Laue was among the few Germans with whom he still felt a common bond. You can probably recall Professor von Laue.

HW: I know his name well, and I can also remember how he looked, but I can recall nothing else about him today.

FH: You will probably recall much more readily the physician Professor Plesch, whom you have mentioned on several occasions.[17]

HW: I recall meeting him for the first time in the spring of 1928, when Herr Professor suffered from heart disease and was lying in bed. The actual house doctor was *Sanitätsrat* Dr. Juliusburger, and later the internist Professor Ehrmann was called in for consultations. And then came Professor Plesch. It may be that Professor Ehrmann called him in.

FH: It is just as likely that Plesch offered to treat Einstein on his own, having been acquainted with him for some time already.

János Plesch, born in Budapest and the same age as Einstein, was at the time an associate professor for internal medicine at Berlin University. He worked at the Medical Clinic II under the internist Friedrich Kraus, whom Einstein occasionally consulted in the Charité—as was related by Theodor Brugsch in his memoir, *Physician for Five Decades*. Apart from that, Plesch also conducted a lucrative private practice.

HW: Beginning in 1928, Professor Plesch was a frequent guest in the Haberlandstrasse. He almost always brought a large gift basket, and, on one occasion, also a chicken that I had to cook right away, so that Professor Einstein would have fresh chicken broth. Before Herr Professor's illness in the spring of 1928, I had not heard of Plesch; he appeared only during the illness. Later, he used to provide his car for the drive to Caputh.

FH: In point of fact, one of his cars—for Plesch owned several. That fact emerges from a report about his emigration, which took him initially to Switzerland, and later he lived in England.

HW: I did not know that.

FH: Judging from photos, he seems very elegant and dressed according to the latest fashion.

HW: That he was, and he was indeed a strikingly good-looking gentleman, as they say. He advised me once to write down everything that I experienced and observed at the Einsteins, and that I could later make a lot of money with it. That is literally what he said to me; I remember it very clearly.

FH: Unfortunately, you failed to follow his good advice. That would now be a great help. If you had kept a diary and had recorded all the personages who visited Einstein, it would be a treasure chest for documenting Einstein's life during his last years in Berlin.

HW: Yes, I also regret that I did not make notes at that time. When Herr Professor was sick, a Japanese gentleman came who barely dared to enter. He bowed many, many times as he entered the bedroom of Herr Professor, who was in his sick bed. He was tremendously reverential, but who he was I cannot recall today. It might be quite interesting to know that.

FH: In this connection, a question regarding Einstein's fiftieth birthday on March 14, 1929. Biographies tell us that Einstein left Berlin on that day in order to escape the onslaught of well-wishers. Supposedly, he spent the day in Gatow at an estate belonging to a friend. Frank refers to that friend as a "boot polish manufacturer," but that must be erroneous. It was Plesch who owned the Gatow estate where Einstein repeatedly sought refuge in 1928 and 1929 so he could work without interruptions.

HW: Where Herr Professor spent his fiftieth birthday, I cannot be certain of. I only know that he was not in the apartment.

FH: In early 1929, Einstein wrote from Gatow to the engineer Michele

Besso, his friend and former colleague at the patent office in Bern. He wrote that from time to time he was now staying all alone at a rural estate where he cooked for himself "like the ancient Eremites." There he had noticed, to his astonishment, how nice and long each day was and "how superfluous were most of the bustling, pointless pursuits that one is normally yoked to." This letter, from January 5, 1929, suggests that in the middle of March 1929, Einstein was again in Gatow. Were you ever there, perhaps to bring something to the "Eremite"?

HW: No, I was never there. But the fiftieth birthday of Herr Professor I recall very well. Fräulein Dukas and I were in the Haberlandstrasse to receive the birthday mail. There were many letters and telegrams and masses of flowers that filled the guest room, because in the other rooms, there was no more room for the many presents.

FH: Some years ago, you showed me the thank-you poem that Einstein wrote on March 14, 1929, and sent to his well-wishers in a facsimile of his hand-writing. It ended with the lines:

> Now that the day draws to its end,
> I send you all my compliment.
> All you did, was done in style
> And now the sun begins to smile.
>
> *Nun der Tag sich naht dem End'*
> *Mach ich Euch mein Kompliment*
> *Alles habt Ihr gut gemacht*
> *Und die liebe Sonne lacht.*

We might now ask if the sun really did smile in Gatow at the time Einstein composed this poem, or if he wrote the last line merely for the sake of the rhyme. Einstein's pronouncements have been put to the test by astronomers and physicists and were—mostly—confirmed by them. Why not meteorologists? I enquired at the Bureau for Climatology in Potsdam about the weather in the Potsdam region on March 14, 1929, and obtained this response: "It was foggy until noon, when it turned hazy, and from 2:45 p.m. till 5 o'clock there was sunshine." Since Einstein composed his thank-you poem toward evening, the line about the sun smiling spoke the honest truth—unlike some biographers. After this bit of fun, let us return to Einstein's guests in the Haberlandstrasse.

HW: Among the medical men who were frequent visitors of Herr Professor, I would put Professor Katzenstein in first place. Although he was not actually the house doctor, he was almost part of the family. He and Herr Professor had a very warm friendship.

Moritz Katzenstein (1872–1932) was an associate professor of surgery at Berlin University (Charité hospital) and his principal position was that of chief physician at the Friedrichshain Hospital. His friendship with Einstein began in the early twenties, and Einstein has acknowledged that, in his Berlin years, there were very few who were as close to him as he. In the beautiful obituary that Einstein dedicated to this meritorious physician and researcher, he says: "We both sensed that this friendship was blissful, not only because each of us understood the other, was enriched by him, and found in him the resonance that is so indispensable for any truly alive person; this friendship also helped us both to be more independent of our outside experiences . . ." Einstein admired in Katzenstein, above all, his highly developed sense of responsibility and his manifest commitment as a physician. He added: "As often happens, he had performed several hazardous operations that morning, and just before we went on board, he enquired by telephone about the condition of certain patients that worried him; I observed how deeply he was affected by the fate of those entrusted to him."[18]

Among Katzenstein's medical research work, Einstein was particularly interested in his animal experiments and his investigation of tissues grown in nutrient solutions. Einstein's obituary closes with these words: "I, however, am grateful for the good fortune that allowed me to be a friend of this kind, indefatigable, and highly creative man."

HW: It was my impression that Professor Katzenstein was indeed an extremely close friend of Herr Professor, far closer than Professor Plesch. Since we are talking of physicians, I would also like to mention Professor

Bucky. He also showed up during the last few years, when Frau Dr. Kayser [Einstein's stepdaughter Ilse] was very ill and he was called in. As I recall, he was a specialist in X-ray therapy.

Gustav Bucky was the originator of X-ray therapy, the therapeutic use of soft (i.e., long wavelength) X-rays. He invented a honeycomb grid, named after him, that improved the existing X-ray sources, and he was the chief radiologist in several Berlin hospitals. Later, after his emigration, he became Einstein's house doctor, even though his home and his practice were in New York, almost a hundred kilometers from Princeton. In his memoir, Bucky speaks of the "magical power" that emanated from Einstein's personality and affected everyone he encountered. Bucky also praises the courageous stand Einstein took on behalf of all oppressed people, in public and also privately.

FH: I made a note of a few scientists who were possible visitors of Einstein. First, Walther Nernst, a physical chemist and a Nobel Prize winner in chemistry.[19] In the spring of 1913, he, together with Max Planck, had sought out Einstein in Zurich in order to win him for Berlin. With the support of the physicists Heinrich Rubens and Emil Warburg, Nernst and Planck had pushed through Einstein's appointment to the Prussian Academy of Sciences; and thanks to his close relationship with a public-spirited banker, Nernst had made it possible for the Academy to offer an enticing salary to the professor in Zurich. Nernst considered his success in bringing Einstein to Berlin one of his greatest accomplishments.

HW: The name is not familiar to me. It seems unlikely that this scientist belonged to Herr Professor's inner circle of friends; otherwise I would remember him.

FH: But you can probably recall Fritz Haber, also a chemist and Nobel Prize winner.[20] His synthesis of ammonia earned him immense scientific recognition, but during the First World War he invented chemical weapons and supervised their deployment at the front. Did you hear his name?

HW: I remember his name very well, but I cannot say with certainty that he was a frequent visitor.

FH: Well, you should have heeded the advice of the handsome János and kept a diary, along with a sort of attendance record. That would be a wonderful memory aid now.

HW: If I would still have that notebook! I dislike keeping written materials, including letters.

FH: Now to the artists among Einstein's visitors. You have told me that Heinrich Mann was in the Haberlandstrasse.[21]

HW: I think it is more likely that it was Thomas Mann.[22] But it might well have been Heinrich Mann, because he lived very close by. I learned that from a book about Heinrich and Nelly Mann that I read recently.

FH: That book, *Farewell to Europe*, also contains a truly devastating judgment of Einstein's violin playing, which does not stem from Heinrich Mann, however: that his playing was "pathetic."

HW: I really laughed out loud when I read that.

FH: Later on, we will talk at greater length about the conflicting assessments by contemporaries of Einstein's violin playing. But first, something about his relation with Thomas Mann. In Germany, there was no close connection between these two Nobel Prize winners, apart from accidental encounters. In her reminiscences, Katia Mann mentions Einstein only fleetingly as a neighbor in Princeton (i.e., following their emigration). She says that he had "such big goggle-eyes" and was very congenial but not particularly stimulating, and that he was not a very impressive personality in daily life. What do you say to that?

HW: I would not subscribe to that. From where I stand, Herr Professor was most impressive in his daily life. And others have also said as much!

FH: To be sure, Einstein could not compete in appearance with a representative of the upper bourgeoisie like Thomas Mann, nor would he have set great store by it. Besides, before Thomas Mann emigrated, he and his family lived not in Berlin but in Munich, so that it seems unlikely that the two had close relations with each other in Germany. Of course, this does not exclude an occasional visit in the Haberlandstrasse by the famous novelist.

HW: I am certain that Thomas Mann was among his visitors.

FH: Einstein had demonstrably close relations with another author who had been crowned with the Nobel Prize. You testified during your interrogation by the criminal investigation police that Gerhart Hauptmann was a frequent visitor in the Haberlandstrasse.[23] According to the literature, Einstein had first met the author of *The Weavers* on Hiddensee Island, where they both used to spend their summer vacations.

HW: Gerhart Hauptmann visited often with his second wife, Margarete, and with his son Benvenuto. I remember that very clearly.

FH: Where did the conversations with Hauptmann and his companions take place?

HW: Mostly in the living room. But often the Hauptmanns also sat at the small table in the dining room—I can see it clearly before me.

FH: You told me years ago that the Hedwig Wangel was a frequent visitor in the Haberlandstrasse and that she always smoked cigars.[24]

HW: She was the only woman I ever saw smoking cigars. That is probably why I can recall her so distinctly.

FH: Hedwig was famous for having interrupted her successful stage career with Reinhardt to devote herself to social welfare, some of the time in the ranks of the Salvation Army. Later on, she returned to the theater, and, at the time of her visits at the end of the 1920s, she was again active on the stage as a character actress. I read that her starring stage roles included that of the nurse in Shakespeare's *Romeo and Juliet* and that of Marthe Schwertlein in Goethe's *Faust*.

HW: Unfortunately, I never saw Hedwig Wangel on the stage.

FH: In the police transcript of September 5, 1934, you named, apart from Max Planck and Gerhart Hauptmann, the music director Erich Kleiber as one of Einstein's visitors.

HW: He was a frequent visitor.

FH: Did they play together?

HW: That I cannot recall, but it is quite possible. I was not always at home, after all.

FH: There exists a photo that shows Einstein with the pianist Joseph Schwarz and his son in front of the piano in the Biedermeier room. At the bottom of the photo, Einstein had written this dedication:

> To the father and his lad,
> The music-making wasn't bad.

Dem Vater und dem Sohne.
Das Spielen war nicht ohne.

HW: The name Joseph Schwarz does not ring a bell.[25]

FH: It must have been before your time.

HW: I would also like to mention Frau Katzenellenbogen among the Einsteins' circle of friends. Estella Katzenellenbogen. Her name popped up in my mind again yesterday, and I immediately made a note to bring it up during our conversation. She was a frequent visitor and, as far as I know, she owned a number of flower shops. She also often put her automobile at Herr Professor's disposal, to drive him to concerts or the theater, or, later, to Caputh. It was a beautiful limousine, much more elegant than Professor Katzenstein's car, for example. Her shops evidently made her very wealthy.

FH: It strikes me that there were very many wealthy persons among Einstein's friends and acquaintances: Frau Toni Mendel with her splendid villa on the Wannsee, Professor Plesch with his lucrative private practice in his sumptuous house on the Budapester Strasse, and now Frau Katzenellenbogen with her profitable flower shops—and they all had shiny chauffeured limousines. But there were also visitors who came by public transportation; for instance, Planck, or the astronomer Archenhold whom I mentioned at the beginning of this conversation. Do you remember him?

HW: Dr. Archenhold was there very often. I would count him also among Herr Professor's closer friends. He often gave me free tickets for talks at the observatory. Dr. Archenhold's demeanor was very modest, and his appearance was not particularly elegant.

FH: Did any orthodox Jews visit Einstein, with side locks and wearing kaftans, as is customary for practicing Jews?

HW: That happened occasionally, but not often. They generally drank only tea, and at most they ate a biscuit, maybe a piece of cake. Mostly, however, they ate nothing at all because it just wasn't a kosher household.

FH: You mentioned that among the scientists there was also Dr. Walter Mayer, who was then Einstein's assistant.

HW: I knew him very well. But he was not there when I joined the household; he came later, perhaps in 1928 or 1929.

FH: When did he work with Einstein?

HW: As far as I recall, he came pretty regularly, and usually he immediately went upstairs to Herr Professor's study, the tower room. From what I heard, he emigrated to America in 1933. He made a very sympathetic impression on me.

FH: According to one Einstein biography, the family used to call him "Mayerle."

HW: That form of address I cannot recall.

FH: Dr. Mayer came from Austria. He was preceded by a young Hungarian mathematician, Dr. Cornelius Lanczos, whom I met in 1965 at the Einstein Symposium of our Academy of Sciences. He was then a white-haired scholar and professor at a research institute in Dublin, and during the intermissions we had very animated conversations about Einstein. He has died since then.

HW: His name is not familiar to me.

FH: That may well be, because soon after you joined the household, Dr. Mayer arrived and remained Einstein's "calculator" until the end. Einstein always had to have a mathematical collaborator because he was—by his own admission—"not a calculator." In his Berlin time, he had altogether three such "calculators," the last one being Dr. Mayer.

But now to a rare visitor, one that had nothing to do with mathematics, and probably also understood little of it, but one that you will surely recall: Charlie Chaplin. He was among the most famous film actors of that time and visited Einstein in 1931. He wrote about it in his memoir, which I have mentioned before.

HW: Yes, he was there. That I know, but he was only at the Haberlandstrasse, not in Caputh.

FH: What was your impression of him?

HW: His appearance actually disappointed me a little. From his films, I had imagined him differently. He was small and dainty. His visit had been announced beforehand.

FH: Were there also visitors who came without announcing themselves?

HW: Yes, some came totally unannounced, and they were mostly closer friends of Herr Professor, who were not served anything. But most visitors did announce themselves. When they telephoned, I had to transfer the call or ask if the time proposed for the visit was convenient.

FH: There is one foreign visitor whom you probably remember very well: the Indian poet-philosopher Rabindranath Tagore, a winner of the Nobel Prize for literature.[26]

HW: I remember him very well, but he was only out there in Caputh, not in the Haberlandstrasse. His visit, naturally, caused a sensation in Caputh, if for no other reason than that he was dressed in the Indian manner. He was tall and slender, a striking figure.

FH: Did students occasionally come to consult with Einstein in his apartment or in Caputh? He did lecture at the University, after all, though he had no office there.

HW: That was certainly not the case; otherwise I would have noticed it.

FH: Were there other visitors that you retain in your memory? You have already mentioned Frau Mileva, Einstein's first wife, who sometimes came to Berlin from Switzerland.

HW: Yes, but only rarely. Herr Professor's sister, on the other hand, whom everyone called "Aunt Maja," came often from Italy. She always brought some Italian delicacies with her and showed me how to prepare them—for example, risotto. That is rice sautéed with onion and garlic in oil before it is cooked. Dear Aunt Maja showed me precisely how to do all this. She came to the Haberlandstrasse, as well as to Caputh. In Berlin, she slept in the guest room, and she often stayed for quite a long time.

FH: It was only recently discovered that at the beginning of the century Maria Einstein studied Romance philology for several semesters at Berlin University. Since women were then not allowed to study at Prussian universities, she was not registered. She had come to Berlin from Switzerland in order to attend the lectures of Professor Adolf Tobler, a celebrated scholar of the language and literature of Old French, and to work with him. "Aunt Maja" was therefore familiar with Berlin from before.

HW: I remember her very well. After the war, she also sent me greetings through Margot, from Princeton, where she stayed with her brother. She was very ill and was unable to return to her husband in Italy.

FH: Maria Winteler-Einstein, PhD, died in 1951 in Princeton. We learn from Einstein's letters to his old friend Solovine that in her last years Einstein sat at her bedside almost every night and read to her from the works of the Greek natural philosophers and the eighteenth-century French materialists, and that her death had devastated him. But back to Berlin. Did close relatives of Frau Elsa also come to visit?

HW: Yes, Aunt Paula, called "Tante Paulchen," and Aunt Hermine. They were both sisters of Frau Professor. Aunt Paula was a sort of favorite aunt of the family, more so than Aunt Hermine, for whom being "genteel" was important. Tante Paulchen, on the other hand, was completely artless and modest.

FH: Where did the two aunts live?

HW: In Berlin. They visited the Haberlandstrasse frequently, and they often brought their relatives with them.

FH: Can you recall other relatives as visitors to the city apartment or in Caputh?

HW: "Tante Maja" from Italy—as I recall, she lived in the vicinity of Florence—and the two aunts in Berlin were, apart from Herr Professor's two sons, the only close relatives that I recall visiting. But there are two out-of-town visitors I don't want to omit, two scientists: the Professors Ehrenfest and Ehrenhaft.

FH: Paul Ehrenfest was a theoretical physicist at the University of Leiden in Holland, and Felix Ehrenhaft was an experimental physicist at the University of Vienna.

HW: I always confuse the two names, Ehrenfest and Ehrenhaft.

FH: Well, that is easy enough to do.

HW: Whenever Herr Professor went to Leiden for his guest lectures, he always stayed with Ehrenfest. One time, upon his return from the journey, he told of sitting in an armchair in Professor Ehrenfest's home and laughing so uproariously that he slapped the upholstered arm rests so hard that clouds of dust emerged from the upholstery and swirled around the room. Herr Professor said that he was quite embarrassed, but Professor Ehrenfest surely did not hold it against him. He was a very cheerful man and loved to laugh, just like Herr Professor. You could hear their laughter all over the apartment.

FH: That reminds me of a quote of Galileo's in Brecht's play: "The theologians have their bell-ringing and the physicists have their laughter." János Plesch also wrote this about Einstein in his autobiography: "Laughter is one of the greatest gifts the Gods bestowed on him. He can laugh uproariously about jokes or comical situations. He also laughs, and that is odd, when others are crying. I have heard him laugh loudly over matters that affected him very deeply." This might well be an accurate description, at least for the time he lived in Berlin.

HW: Yes, that is exactly the way Herr Professor was.

Paul Ehrenfest was, according to Plesch, "a splendid example of old-fashioned Austrian warmth." He came from Vienna and had married a Russian physicist. In September 1933, he tragically put an end to his own life. According to Plesch, the immediate reason was the fact that his son was in danger of becoming blind. In Einstein's obituary for his friend, he gives other reasons for his desperate act, among them a growing estrangement between Ehrenfest and his wife, Tatiana. Einstein writes that this was "a terrible experience for Ehrenfest with which his wounded soul was unable to cope." In Einstein's opinion, Ehrenfest also suffered from the fact that his critical aptitude exceeded his creative capability. It seems therefore that a combination of several motives led to a death that is so at odds with that cheerful, always heartily laughing scientist. Tatiana Ehrenfest survived her husband by over thirty years and died in Leiden in 1964.

FH: Speaking of laughing, some time ago you told me of a funny misunderstanding that happened to Einstein on a train journey in England. It must have been around 1930. Would you please tell it again?

HW: With pleasure. I don't know who was sitting in the train with Herr Professor, but when the train passed through a town with a great many smoke stacks, his companion said to Einstein: "*Das englische Essen*," referring no doubt to the town's great armament works.[27] Herr Professor was, however, deeply immersed in his own thoughts, and responded with: "Yes, it is ghastly. They cook everything with mutton fat." This misunderstanding cracked both of them up and when Herr Professor told the story in Berlin, in my presence, he was again convulsed with raucous laughter. He told it during the midday meal—that had been prepared very differently than in England. The cooking there was not to his taste.

FH: You mentioned the Viennese experimental physicist Ehrenhaft a while ago. Do you remember him?

HW: Yes, indeed, very well, because one summer the Ehrenhaft couple lived in Caputh for about three weeks. They lived upstairs, in what was otherwise the Kaysers' room. They were both extremely kind and amiable in the Viennese manner, and we often chatted together. They were very loquacious and not at all restrained toward me, a domestic servant.

In 1908, Felix Ehrenhaft became known for his interesting experiments with electrons. He believed he had shown that electrons were not elementary particles in the strict sense, but were divisible, somewhat like atoms, so that "sub-electrons" therefore had to exist. That view led to lively discussions among physicists of that time. The often-cited quotation from Lenin's philosophical work *Materialism and Empiriocriticism*, "The electron is just as unfathomable as the atom," was written at that time and was seemingly based on these considerations. At present, the issue of "sub-electrons" is again under discussion, particularly among American physicists.

FH: Einstein was a close friend of Ehrenhaft, as well as of Ehrenfest. He stayed in Ehrenhaft's home whenever he was in Vienna, and Philipp Frank relates in his biography how Frau Ehrenhaft had once taken the extra pair of trousers that Einstein had brought for his evening lecture to the tailor to have them pressed. Later on, she noticed with dismay that her guest nevertheless left for his lecture in the unpressed pair, still badly wrinkled from the train journey. Trouser creases just weren't important for Einstein.

HW: No, that was not his strong suit. Later on, after his step-son-in-law Dr. Marianoff moved in, *his* suits were picked up by the tailor to be ironed every four weeks. That was completely foreign to Herr Professor; it was not habitual for him. Only very occasionally, when it was absolutely necessary, was one of his suits pressed. Now and then we did bring something to the tailor around the corner, but only the good suits for going out.

FH: So Plesch was on the right track when he wrote about Einstein, "Outward appearances meant nothing to him. He was most comfortable in a leisure suit and slippers, or, in summer, in sandals, linen pants, and a jersey. Dressed like that he would go sailing, alone, for hours; and if he needed a head covering, he took his handkerchief and tied knots in all four corners, so snugly that he could stretch it over his entire skull."

HW: Yes, that's the way it was.

FH: In János Plesch's memoir there is also an example of how bluntly Einstein could express himself. We have already mentioned that in 1928 or 1929, before the summer house in Caputh existed, Einstein occasionally stayed on Plesch's estate in Gatow, so that he could work in peace. Not far from the estate was a sewage farm and, when the wind blew in the right direction, the smell was palpable on the estate. The sewage fields ought to have been relocated long before, for reasons of hygiene, but the city council of Greater Berlin had delayed the measure again and again. One day, when Berlin's mayor [Gustav] Böss was dining at Plesch's house and Einstein was also present, Böss asked Einstein, a little guiltily and embarrassed, if the smell did not bother him? Whereupon Einstein answered: "It doesn't bother me particularly, and from time to time, I return the favor."

I consider this account to be credible. Einstein could be very ironic; that he could also be quite outspoken is shown by his response to an American women's organization that protested against his return to the United States, in the fall of 1932. This is the way Einstein ended his response: "But are they not right, these watchful citizens? Why invite to my home a person who devours hard-nosed capitalists with the same appetite and pleasure, as long ago in Crete, the monster Minotaur devoured Greek virgins; a person who is, furthermore, vulgar enough to reject all wars—excepting the inevitable war with one's own spouse? Therefore pay close attention to your clever and patriotic womenfolk and bear in mind that Mighty Rome's capital city was once saved by the cackle of its loyal geese!"[28]

That is the end of the citation. The remark about the unavoidability of war with one's spouse must be seen as a generalization of his own marital experience, specifically, during his last years in Berlin, and our conversations have hinted at this. But one ought not take this insight of Einstein—wholly in his spirit—too seriously.

THIRD CONVERSATION:
FAMILY—VACATIONS—FOREIGN JOURNEYS

FH: I have a question about Einstein's summer vacations before there was a country house in Caputh. You arrived at the Haberlandstrasse in June 1927. Can you recall where Einstein spent his summer vacation that year? It is known from his letters that he was in Switzerland in July. Did he go to the seashore afterward?

HW: I cannot say anything about that. I only know that, following his serious illness in the spring of 1928, Herr Professor stayed at the seaside. Unfortunately, I don't remember the name of the place, only that it was a small fishing village, not an elegant resort or the like.[29] The whole family went there for three or four weeks.

FH: There is a photo that may well be from that time that shows Einstein on a deck chair, wrapped in blankets.

HW: I remember that photo well; I saw it in an illustrated magazine. It was taken in that small seaside village.

FH: Were you there with them?

HW: No, I took my vacation during that time. I generally took my vacation when the Einsteins were away. I would then stay with my parents in Lautawerk. Sometimes Margot remained in Berlin, and in that case I naturally continued to run the household.

FH: Einstein, supposedly, often vacationed on the Baltic shore, in Ahrenshoop and on Hiddensee [Island].

HW: In my days, only the daughters, Frau Dr. Kayser and Margot, went to Hiddensee, where they stayed in Kloster. Herr Professor no longer went there, that must have been at an earlier time.

FH: After the Einsteins moved into their summer house in Caputh, did they still take their vacation on the seashore, or elsewhere?

HW: No, then Herr Professor always remained in Caputh. He liked it very much there. Of course, he often left for lectures or scientific meetings, and so on. Once, I remember, he telephoned Frau Professor from Rothenburg ob der Tauber and asked her to join him and see that beautiful small mediaeval town. She did indeed go and they stayed there for several days. It must have been in early spring.

FH: Where did the two daughters spend the summer in the last few years in Berlin? Did they still go to Hiddensee, or did they come out to Caputh?

HW: They stayed in Caputh afterward. In the winter, they often went to the Engadin, but they spent their summers mostly in Caputh. When I first came, they often went to Hiddensee, particularly Margot. She told me once of running into Gerhart Hauptmann's second wife, Margarete, there, whom she had met before. When Margot told her that she now had a parakeet, a sweet little bird, Frau Hauptmann replied maliciously, "Oh, Margot, you have had it forever, already." She was quite outraged when she told me about it. The two women could not stand each other.

FH: The parakeet must have been much loved by the family and must have been the center of attention in the city.

HW: Yes, that parakeet, a blue one, was really a dear creature. They were much rarer then, than they are now. Margot still spoke about her "Bibo," or "Biebchen," in her letters after the war.

FH: I read in a biography that the parakeet had been presented to Einstein by a Japanese scientist.

HW: That is news to me, I don't know that. I only know that Bibo could speak very well and that Margot had taught it to him.

FH: Did the Einsteins keep any other animals in the Haberlandstrasse or in Caputh?

HW: In the Haberlandstrasse only the parakeet, which was greatly enjoyed by Herr Professor, but most of all by Margot. His cage stood in her room, but he was often free to fly all over the apartment. But wait a moment—I just recalled that on the windowsill in the Biedermeier room stood a rectangular aquarium with one or two goldfishes; it was very simple, not like today's.

FH: Did Einstein take an interest in the goldfishes?

HW: Herr Professor occasionally sprinkled some fish food for them. There were no other animals in the house; only later, in Caputh, we had a pet dog, but not one we owned, and in the last year, a tomcat that adopted us. That's all.

FH: Once a week, when Einstein was not at home, you went up to the tower room to clean up.

HW: Yes, and I used to wash the windows at the same time, something I didn't have to do downstairs, but no stranger was supposed to enter Herr Professor's study.

FH: Were you allowed to disturb the books or papers on Einstein's desk when cleaning up, or did everything have to remain as it was? Many scholars are very particular in that regard and want nothing to be touched or changed, fearful that they would no longer be able to find anything.

HW: Actually, everything was always quite orderly. Besides, Fräulein Dukas also used to work upstairs with the typewriter, although not every day. I often brought tea to her upstairs; there was no dictation or typing in the apartment downstairs, except, perhaps, when Herr Professor was sick in bed. That could be.

FH: Did Einstein use the typewriter himself? Many scientists do, using their own original fingering with the right hand not knowing what the left is doing.

HW: I never saw or heard Herr Professor typing.

FH: Did Einstein, who was, after all, interested in technical advances, use a dictating machine? These devices were already around then and were being improved continually. Electronic dictation machines, like today's, did not exist, of course. They were mechanical devices that used wax cylinders and were based on the same principle as Edison's phonograph. One had to speak into a funnel, and the recordings could be erased with an erasing device. It was all somewhat complicated, but it served its purpose. The chemist and social scientist Wilhelm Ostwald acquired no less than three dictating machines, so that he could manage his voluminous literary output. Did Einstein have such a machine?

HW: No. He dictated directly to Fräulein Dukas's shorthand book or, sometimes, to her at the typewriter. Apart from that, he wrote everything by hand.

FH: One more question about his work habits. As you know, a while ago there was a dispute covered by the daily press, regarding some shorthand pages supposedly written by Einstein. Well, it is not impossible that Einstein knew shorthand: he had an opportunity to learn it in the cantonal school in Aarau where he spent his final high school year before attending university in Zurich. It is known that other physicists of his generation, for instance, Lise Meitner and Erwin Schrödinger, used the Gabelsberger shorthand method for making hasty notes. Did you ever notice papers with what looked like shorthand script on his desk in the tower room, or downstairs laying on his night table in the morning?

HW: I saw many pages and notes with mathematical formulas that Herr Professor had written, but I never noticed any shorthand script.

FH: I was able to clear up that matter by comparing handwriting samples—from documents provided by the Arnold Zweig Archive in Berlin. The purported Einstein shorthand manuscript was actually written by the painter, graphic artist, and author Hermann Struck, a Zionist who was a host of Einstein when he visited Palestine on his return journey from Japan in 1923.[30]

HW: I read about that in the newspaper.

FH: You told me some time ago that there were no games at the Einsteins, neither Bridge nor Skat nor Halma nor dominoes or the like. Einstein, for his part, despised the game of chess, even though he was a friend of the chess world master Emanuel Lasker, with whom he often discussed philosophical and other matters. In 1952, he wrote in a preface to a Lasker biography that the game's "power struggle and its competitive spirit had always repulsed him." Earlier, in 1936, he had already expressed his sentiment regarding, or rather, against chess in a major American newspaper. He rejected this ingenious game on moral grounds, because it is based on outwitting the partner by all sorts of dodges. So you would hardly have expected a chess table, usually an inlaid one, or a chessboard in the apartment.

HW: No, I don't remember anything like that.

FH: That answer was predictable, I just wanted to make sure. An American biography states that in Princeton a framed diploma of Einstein's hung on the wall, one of the many honorary diplomas he received in his scientific career. It certified his membership in the Swiss Natural History Society in Bern. Can you recall such a framed certificate on the wall in the Haberlandstrasse?

HW: A diploma like this hung in Herr Professor's bedroom, in the corner next to the door, where the silver chest stood. I remember it with certainty; it may well have been that diploma.

FH: Did you see the diploma for the Nobel Prize, or the gold medallion that comes with it?

HW: Never, unfortunately. I assume that Frau Professor took care of all these medals and diplomas. Herr Professor cared very little about those things.

FH: That is most likely true, for one biography stated that Einstein apparently did not know what the Nobel Prize medallion looks like.

HW: That is quite possible.

FH: What was the situation with regard to Einstein's smoking, in your time? He had been a passionate smoker since his student days, and he remained so into his advanced age. He finally had to give it up in Princeton, on doctor's orders. Afterward he would only "smoke cold" [i.e., without lighting the pipe]—as they say. What about the last years in Berlin?

HW: Herr Professor smoked a lot, mostly a pipe.

FH: Max von Laue, who visited the father of relativity theory at the patent office in Bern in 1906, reported that on their walk through the town, Einstein offered him a stogie that tasted so bad that he "accidentally" let it drop from a bridge into the Aare River.[31] In your time, did Einstein smoke, apart from his pipe, stogies or other cigars, or maybe cigarettes?

HW: No cigarettes, but Herr Professor did occasionally smoke cigars, particularly when there were visitors.

FH: Plesch writes that for Einstein lighting a cigar was a "ceremony" that his senses savored and that he gave it a "festive connotation." He continued: "Officially his concerned wife allowed him only one cigar a day, but surreptitiously he kept in his redoubt many boxes of cigars of the most diverse brands that friends had smuggled in, using cunning ruses." The tower room was surely his redoubt. When you went up there to bring tea or the like, was it always thick with smoke?

HW: Yes, indeed, there was always a blue haze. However, Herr Professor smoked mostly his short pipe there.

FH: Can you tell me something about Einstein's height? There are quite variable reports. Fräulein Dukas described him as being a "little over median height."

HW: Herr Professor was a little taller than I, and I was then 1.68 m in height. He might have been 1.72 or 1.74 m [5ft 8in].

FH: A Swiss acquaintance from his youthful days gave his height as 1.76 m.

HW: Well, I don't recollect him as being that tall.

FH: Since we are talking about Einstein's appearance, I would like to raise an issue that I found in Plesch that seems remarkable to me. Plesch writes that to the astonishment and annoyance of graphic artists, painters, and sculptors, the back of Einstein's head was missing, and that is why hardly any artist had caught his likeness. He thought they must all have been inhibited when faced with this enormous skull without a back. There must be some truth to this.

HW: That is something I never noticed. What I do know is the Herr Professor was rather broad-shouldered, the opposite to Geheimrat Planck, who was a little smaller than he and quite delicate. Professor von Laue was also not as broad, I would say.

FH: A psychological question: How would you classify Einstein according to the familiar four temperaments? Was he choleric, sanguine, phlegmatic, or melancholy?

HW: I would say that Herr Professor was calm, but I would not call him phlegmatic. Choleric, irascible, he was not. Nor was he one of those people who are quickly driven up the wall. I, at least, never experienced him like that.

FH: This surely applies to the family circle, with occasional exceptions for the rather loud arguments that you have spoken of. It is reported by outsiders, however, that Einstein could also come across as being brusque and forceful. He was simply full of contradictions. Plesch claims that Einstein was capable of hating and deep loathing; he writes, "It is difficult to become his enemy, but once he had expelled someone from his heart, he was finished with him for good." I think this is a fitting characterization. In the same spirit, Einstein's colleague and biographer David Reichinstein wrote, "Einstein can express strong dislikes, he can become very vehement, impatient, even unjust." It was not impossible to rouse him out of his calm, as is shown, by his behavior later on, in particular.[32] I have expanded on this and provided sources in several sections of my book *Einstein and His World View*.[33] Einstein was not always justice incarnate.

HW: But Herr Professor was not moody, and most of all he was always very polite to visitors. Also to relatives of mine when they came to visit; he always greeted them warmly and shook hands with each of them. There was never anything like: "Here is Herr Professor, and there is a mere domestic servant." He would always address my cousin, who visited me often, as "Frau Hedwig." I have already told you how Herr Professor telephoned Professor Katzenstein to make sure that Hedwig's six year-old boy would receive particularly good care in the Friedrichshain Hospital.

FH: Some biographies leave the impression that Einstein was rather careless in his private life, even slovenly. It can hardly be denied that he paid scant attention to keeping his papers and correspondence in order. Max von Laue with whom I had many conversations about Einstein, told me literally, "Einstein was not a collector." Apparently, it was only Helen Dukas who put his files in order, so that from the time before 1928, much has been irretrievably lost, such as the correspondence with other scientists, which would be critical for gaining a deeper understanding of how his theories were created. Einstein was indeed not a collector, but could one call him slovenly?

HW: I would not say so. I would prefer to call his conduct indifferent. For example, as we mentioned before, it was all the same to him what he wore; it should just be as comfortable as possible.

FH: Since you brought clothing up again, I would like to add this: A biography has an account from about 1925 of Einstein walking through the Tiergarten Park accompanied by a student, on his way to the University. It had rained heavily earlier, and there were many puddles in the paths. Supposedly, Einstein skirted each puddle carefully and explained that his shoes were really in need of new soles. Do you consider this story credible, exaggerated, or pure invention? There are, after all, so many such anecdotes about Einstein.

HW: I would not consider this an exaggeration or a complete invention. It could well have happened. Herr Professor would put on shoes that he was comfortable in, even if they were no longer watertight. He would wear them until it was no longer possible.

FH: In that case, Antonina may not have exaggerated when she writes that Einstein preferred an oft-mended sweater or an old jacket to wearing a material whose touch was unfamiliar; a dressing gown full of holes was more comfortable than a new one that he received as a present. This characterization evidently applied to his footwear, as well. This reminds me: you once mentioned to me that Einstein liked to wear shoes with crepe soles.

HW: Herr Professor liked to wear such shoes, but he also wore sandals a lot.

FH: Biographers have also reported that Einstein disliked wearing socks and sometimes did entirely without them. He himself wrote from Princeton to his friend Dr. Hans Mühsam in Haifa, the brother of the murdered antifascist Erich Mühsam, that he had become "a lonesome old chap, a kind of old-fashioned figure, known mainly for not utilizing socks."[34] Did he already have an aversion to the use of socks in his last Berlin years?

HW: In Caputh, certainly. In fact there were occasions when he went barefoot, wearing only sandals, to lectures at the Potsdam institutes.

FH: The same letter to Hans Mühsam of March 1953 also contains Einstein's definition of a good housewife. I will cite it because I find it funny and fitting: "The good housewife stands halfway between the outright slob and the neat freak." From the reports I have seen, I get the impression that Einstein's first wife, Mileva, approached the first extreme, while Elsa leaned toward the second.

HW: Frau Professor was very concerned with good order, but I would not call her a neat freak.

FH: One question about the delivery of mail: Did the postman come up to the apartment door, or did he leave the mail downstairs with the porter?

HW: He came up in the elevator, rang the doorbell, and handed me the mail. We did not have a mailbox in the door, nor downstairs. The postman always delivered the mail in person, and if no one was at home he slid the mail through a slot in the door. During the summer months, the post office forwarded the mail to Caputh, and the rural mail carrier delivered it. The letters to Herr Professor were sometimes addressed very magnificently, and I recall one letter from Japan that arrived with the address "Professor Albert Einstein, Germany." And they all arrived without delay.

FH: Something similar is reported of Gerhart Hauptmann, who had become world-famous through his dramas, particularly, *The Weavers*. Once he received a letter from America with the short address "Gerhart Hauptmann, Germany." That was sufficient for delivering it. But what happened next with the mail?

HW: Usually I placed the mail on a little cabinet in the hall, or I took it in straightaway and Frau Professor looked through it.

FH: In doing so, did she make a preliminary selection? I mean, did she give her husband only those letters that deserved an answer in her opinion? Did she act as a censor?

HW: That may have been the case, but I cannot say anything about it.

FH: I raised the question because one biography states that the porter brought the day's mail in large baskets upstairs and Frau Elsa then looked through it. She supposedly spent half her day sorting the mail.

HW: As I said before, the postman brought the mail to the apartment door and I took it. Sometimes there was a lot of mail, but there was no need for large baskets to transport it.

FH: Again, one of those exaggerations and flights of fancy that abound in the Einstein literature. But there is an important matter I want to ask you: Which daily paper did Einstein subscribe to? I assume that it was the *Berliner Tageblatt*, which published articles by him. In August 1920, for example, it published his response to the accusations against him, personally, and against the relativity theory that had been raised at a rabble-rousing rally in the *Berliner Philharmonie*.[35]

HW: I cannot say with certainty that Herr Professor subscribed to the *Berliner Tageblatt*; it could just as easily have been the *Vossische Zeitung*.

FH: My guess that it was the *Berliner Tageblatt* is based on a 1920 letter in which Einstein thanks the paper's owner, Rudolf Mosse, for sending a complimentary subscription to Josef Popper-Lynkeus, the Viennese engineer and antimilitarist author, who was in financial straits and could not afford to subscribe.[36] That suggests that he himself was a subscriber, and, in an Einstein biography I read, the *Berliner Tageblatt* is named as his paper.

HW: I cannot be a hundred percent certain that it was the *Berliner Tageblatt*; it seems to me rather that it was the *Vossische Zeitung*.

FH: It might have been the *Vossische Zeitung*, for Einstein had also published a number of popular science articles in that Ullstein newspaper. His first newspaper article following his move to Berlin in April 1914, appeared there under the title "The Relativity Principle." Ah well. Did Einstein read his paper at breakfast, as was customary in many families?

HW: Not at breakfast. I don't think Herr Professor did much newspaper reading, in any case.

FH: One question regarding his literary preferences. Several biographies report that Einstein was particularly fond of the Russian author Dostoyevsky and that he also enjoyed reading in *Don Quixote* by Cervantes.

HW: There is nothing I can say about that. There were books by Dostoyevsky in the library, however, for that is how I got to know that author for the first time. The novel *The Idiot* is the first book by him that I read. I was allowed to take from the library whatever I wanted to read; all the books were freely available to me.

FH: And you made good use of that permission.

HW: Yes indeed, very often. I have always enjoyed reading. I remember that I borrowed *Der Grüne Heinrich* [Green Henry] by Gottfried Keller from there. Goethe and Schiller's works were, naturally, also there.

FH: I read in an otherwise reliable biography that Einstein had a roll top desk (i.e., a desk whose writing surface with everything on it can be locked under the roll top), the kind of desk one often sees at post offices. Was such a desk in the apartment? You would surely have noticed it while dusting the furniture.

HW: A roll top desk? I only remember simple, flat desks downstairs in the apartment, as well as upstairs in the tower room. I know nothing of a roll top desk.

FH: Some questions about the way family members addressed one another: What did Einstein call his wife?

HW: Elsa.

FH: And what did she call him? Some biographers, among them Frank, report that she called him Albertle, in the Swabian manner.

HW: In my presence, she called him only Albert. Albertle—no, I would have noticed that.

FH: What did the two stepdaughters call him?

HW: Father [*Vater*] Albert.

FH: The stepson-in-law Dr. Rudolf Kayser was presumably addressed as Rudi.

HW: Rudi or also Rudolf. Other than that, he was called Bärchen [little bear] in the family circle. That's also what his wife, Ilse, called him. She herself was called Äffchen [little monkey], and Margot was called Häschen [little rabbit]. They used these nicknames among themselves all the time.

FH: It was reported by a visitor that Dimitri Marianoff was called Dima, and that Margot called him by the diminutive "Dimachen." Was that the case? It would fit in with "Hertachen," as you were called by the daughters.

HW: I cannot recall that form of address, but it is not impossible that occasionally Margot called her husband that.

FH: I read in an American Einstein biography that Einstein called Margot "Liebchen." The word is given in German, to convey that it as a literal quote. Can you recall such a tender appellation?

HW: No.

FH: Did Margot, who was a sculptress and is referred to as such in documents, do much modeling in your time?

HW: Yes, in her room. I still have a memento from her, right here on my wardrobe. Let me get it down, so we can both look at it. This little ceramic was made by Margot, and she gave it to me.

FH: As a farewell present, before she had to flee to Paris in early April 1933?

HW: No, right after she finished making it. I was delighted with it. The model for this sculpture was a dancer whose performances Margot often visited. Then she used her imagination to mold this small figure.

FH: One of the most famous dancers at the end of the 1920s was Anna Pavlova, world-famous for her dance "The Dying Swan." At that time, she presented dance evenings in all European cities, also in Berlin, and one could imagine that she served as Margot's model for the little sculpture. Was her name mentioned?

HW: I know Anna Pavlova's name well, but this little figure was not modeled after Pavlova—that much I know. It represents the dancing of another, not so famous dancer who was still young: Niddy Impekoven. Margot modeled her pieces in clay, and then I took them out to Lichterfelde to have them fired. She made many such pieces, also small animal ceramics. Later on, in America, she wrote me that she studied sculpting in stone.

FH: That is confirmed in a letter Einstein wrote to Max Born in 1937, which reads, "Margot is spending the week in New York and sculpts in stone with unsurpassable enthusiasm."

HW: In her time in Berlin that was too strenuous for her and she made only small things. She had a sort of potter's wheel in her room on which she shaped these pretty pieces.

FH: Did Margot also make heads? For example, the head of the elegant János or other prominent visitors?

HW: No, not that. She did not think she was capable for that; she had only just finished her training, and occasionally she still went to Lichterfelde to see her teacher, Professor Isenstein. She possibly worked on bigger pieces there, but I never got to see them.

FH: Did Margot have friendly relations with famous personalities?

HW: I only know that she was friendly with Frau Hilferding, with Rose Hilferding.

FH: Was she related to the social-democratic economist Rudolf Hilferding, who was twice minister of finance in the Weimar Republic and is the author of *Das Finanzkapital* [Financial Capital]?[37]

HW: Margot's friend was the wife of Rudolf Hilferding, as far as I know. Everyone always only called her Rose.

FH: As you once told me, you were always struck by Margot's sparkling blue eyes and you related a small incident about them. Could you repeat it?

HW: Margot was sitting in a coffee house with some friends—I think it was in Vienna—when the current hit tune was being played: "Where did you get your lovely blue eyes?" [*Wo hast du denn die schönen blauen Augen her?*] While the music was playing, one of her friends asked her, Where did *you* get your lovely blue eyes? To which Margot replied very innocently, "From my mother," which made them all laugh.

FH: From all your accounts, Margot seems like a very sympathetic, girlish young woman, with a lovable, artistic nature. The nickname *Liebchen* that Einstein supposedly used would actually have suited her well, don't you agree?

HW: That may well be, but I just cannot recall that form of address.

FH: I have a few questions regarding the Einstein family's state of health. You mentioned Einstein's rather serious illness in the spring of 1928 several times already. According to Plesch, who as his attending physician should know, Einstein suffered from "acute cardiac stress" [*akute Herzüberanstrengung*]. Plesch called it a "myocardial congestion" [*Myocardstauung*]. Einstein himself, supposedly, admitted that his heart muscle ailment was caused by rowing a heavy sailboat while he was becalmed one evening. That must have been in the summer or autumn of 1927; that is, two years before he was given the *Tümmler*, with its built-in auxiliary motor—in case of doldrums. One must therefore assume that Einstein was using a sailboat without an auxiliary motor when he got into trouble. Other factors may also have contributed to his overexertion, apart from rowing. Plesch mentions Einstein's strenuous hike in heavy snow while carrying a heavy suitcase, when he stayed in Switzerland during the winter of 1927/28. Plesch thought that this might have been too much for even the healthiest person. Later, Einstein wrote to his friend Michele Besso that he had almost "kicked the bucket" [*nahe am Abkratzen war*] in the spring of 1928. And his biographer Carl Seelig reported that when Helen Dukas introduced herself as his private secretary in the middle of April 1928, he received her with the words: "here lies an old corpse" [*Hier liegt eine alte Kindsleich*].

HW: Herr Professor was indeed seriously ill then, and we were all very worried about him.

FH: Was Frau Elsa often ill? She died relatively young, after all, in 1936.

HW: Actually, no, not in Berlin. In any case, I cannot recall Frau Professor being sick for any length of time. After emigrating, she suffered from goiter for a year and a half, before she died. Margot nursed her mother until the end.

FH: It seems that Einstein soon got over her death. In a letter to Max Born he writes: "I have acclimatized myself splendidly here, rummage about like a bear in his den, and I actually feel more at home than at any time in my variable life. This bearishness has been accentuated still further by the death of my mate who was more attached to human beings than I."[38] In his commentary, Born wrote, "It is somewhat strange that Einstein announces the death of his wife, as an aside, while describing his bear-like existence in which he feels at home. In spite of all his kindness, sociability, and his love of humanity, he was nevertheless completely detached from his environment and the human beings that are part of it." In my opinion, other factors also played a role in Einstein's "somewhat strange" conduct, and we will talk about them later. And how was the health of the two stepdaughters?

HW: Margot had a lot of trouble with her gallbladder, and when she had stomach cramps I had to prepare hot-water bottles for her and make her chamomile tea. She would often lie in bed. Frau Dr. Kayser, that is, Ilse, was always sickly and very delicate. She died in Paris soon after emigrating. But Margot was also delicate; both daughters were.

FH: Coming back to Einstein's step-sons-in-law, Kayser and Marianoff: you mentioned in our first conversation that Kayser was an editor at the S. Fischer Verlag. What was your impression of him?

HW: His conduct was simple and modest, and he was very likable. He always came to visit together with his wife.

FH: Did he get along with his brother-in-law Marianoff?

HW: I don't think that they had close contacts with each other. Ilse had already been married for four years when Marianoff came into the house.

FH: Was there a particular celebration of Margot's marriage in December 1930?

HW: Oh, not at all. Everybody went to the registry office, but there was no celebration. Anyway, the parents left immediately afterward for America to lecture in Pasadena. There was no big wedding feast, not at home. It is possible that there was a festive meal in a restaurant, but I know nothing about that.

FH: Dimitri Marianoff published an Einstein biography in English in 1944, just as Rudolf Kayser had done in 1930. Einstein dissociated himself from that book and justified his refutation with the numerous incorrect details in the portrayal. That is indeed true, for the biographical data are inexact or grossly incorrect. But some portions of the biography are valid; for instance, when he writes that "a stout maid, blonde and German-looking" opened the door for him when he visited the Haberlandstrasse for the first time. Judging from your photos from that time, this is a fitting description. Could one tell from Marianoff's pronunciation that he was a Russian?

HW: Yes, he spoke German with an accent.

FH: It is known from a letter Einstein wrote to Max Born that, when Marianoff published his Einstein biography in 1944, he had already been divorced from Margot for seven years.

HW: Margot told me in her very first letter after the war that she was using her previous name again. I would like to add, however, that from the very beginning the relationship between Herr Professor and Dr. Marianoff was not at all like that with Dr. Kayser. It cannot even be compared to it, for it was much more familiar and warmer with Dr. Kayser. I also do not believe that Herr and Frau Professor completely approved of Margot's marriage. In any case, I had the feeling at the time that they were not overjoyed by it. But at first, Margot was very happy, and Marianoff was an elegant gentleman. He may well have had other acquaintances on the side, for after both had fled to Paris bills kept arriving from flower shops for gifts Dr. Marianoff had ordered. They all had to be paid. Toward me he was always kind and considerate, and he often brought me a delicacy, if he passed the Russian shop on his way.

FH: Biographies describe Marianoff as a journalist or author. Was he a freelancer or was he employed by a publisher or press agency?

HW: I can say nothing about that. I only know that he moved a lot in artistic circles. He often gave me tickets to premieres in theaters or in cinemas, and I always made use of them. He always gave me two tickets, so that I could take somebody with me.

FH: You told me years ago that, following their wedding, Marianoff and Margot traveled to the Soviet Union.

HW: Yes, right after their wedding. It might have been both a business trip, and their honeymoon trip, at the same time.

FH: After their return, did the two talk about the impressions they gained on their travels, or show you photos?

HW: I cannot recall seeing photos, but Margot did tell me how beautiful everything was in Moscow but that great disparities still existed. I remember that comment very clearly, and it gave me the impression that they were not only in Moscow but had visited other cities, as well.

FH: Forgive me for changing the subject! I have not asked you yet if, as a domestic servant, you had to wear a special uniform.

HW: In Berlin, I wore my own clothes in the morning, and in the afternoon, for receiving visitors, I had to put on a black dress with a white collar, also a small white apron and a cap, as was customary at the time. When we had a big party, I wore a particularly coquettish little apron, something Frau Professor always set a great store by. All of this was presented to me; I didn't have to get it myself.

FH: And in Caputh? In the afternoon, were you also dressed so formally in a black dress and a white cap? Distinguished guests often came there, as well.

HW: In Caputh it was different. There I used to wear my summer dresses, with a white apron, of course, as was considered proper.

FH: You told me once that Frau Elsa treated you well. Can you give me a specific example?

HW: One year, Frau Professor bought me a long-haired fox [shawl] for Christmas. She meant well, but she must have noticed from my face that the present wasn't to my taste. I told her quite frankly, "I don't care for it; I don't wear this kind of thing." She did not hold that against me but took me to the furrier where she had bought the fox shawl and allowed me to pick out some possum trimmings for my overcoat. That was surely very decent of her, for she might have taken the position: if she is given a fox shawl as a present, she ought to wear it. She also did not blame me for all the little mishaps that occasionally occur in a household. I remember the time when, on the terrace in Caputh, I tried to chase away a wasp and whacked a strawberry cake off the table. I was very embarrassed by this mishap and was annoyed with myself for my clumsiness, but Frau Professor did not reproach me, and they all laughed about it.

FH: And Einstein, naturally, as well.

HW: Herr Professor was not present just then, but when he heard about it later, he laughed loudly and heartily because I had chased the wasp so thoroughly that the whole cake went along.

FH: Then Plesch is quite right when he writes that Einstein could laugh "about the most harmless stuff."

HW: Something just came to my mind that I forgot to mention when we talked of the newspaper subscription. Herr Professor subscribed to the magazine *New Russia* [*Das Neue Russland*]. When I told him that my older brother liked to read that periodical, Herr Professor immediately arranged for it to be sent to him regularly. My brother managed to save two issues through the Nazi period, when he had to hide all such materials. But he kept those two issues and showed them to me recently. One of the covers shows a picture of [Russian author] Maxim Gorky.

FH: This magazine was published by the Society of Friends of the New Russia, and its mission was providing factual information about the cultural and economic currents in the Soviet Union. Some issues had photos showing Einstein at evening lectures of the Friendship Society in Berlin. He is seen standing next to the Soviet geochemist A. J. Fersman and the Soviet Union's long-term minister for education, the art historian A. W. Lunacharsky, with whom he was well acquainted.

In 1923, Einstein was one of the founders of the Society of Friends of the New Russia, and as a member of its board he played an active role in its work; there are photographs that show him sitting with the presidium. Among the members of the society were the writers Alfred Döblin and Thomas Mann, the publishers Samuel Fischer and Ernst Rowohlt, the theater director and producer Leopold Jessner, the painter Max Pechstein, the architect Bruno Taut, the zoologist Julius Schaxel, and the radio engineer Count von Arco. The society therefore represented a wide spectrum of academic professions whose members resisted the anti-Sovietism that was pervasive in the Weimar Republic and increasingly expanded with the rise of Hitler fascism. In 1929, Count Arco, a notable inventor and organizer in radio technology, celebrated his sixtieth birthday in Moscow, in a politically demonstrative manner.

One issue of *New Russia* carries a report of a lecture that Lunacharsky delivered in the Berlin Singakademie, today's Maxim Gorki Theater, in November 1931; it was entitled "The Cultural Successes of the Soviet Union," and the speaker mentioned Einstein's attitude toward the growth of socialism in Soviet lands. He said, "The great physicist Albert Einstein once said to me, very perceptively: First off, I believe in the experiment. I see the reconstruction activity of the Communists in Russia as an experiment on an enormous scale, conducted, in my opinion, under the most unfavorable circumstances in a poorly equipped laboratory. Therefore, should the experiment end in failure, it would not prove to me, as a scientist, the impossibility of success in a better equipped laboratory. A success in Russia, on the other hand, would constitute irrefutable proof of the correctness of the premises it is based on."

Apart from the Society of Friends of the New Russia, Einstein was also active in the Soviet-German society Culture and Technology. He had taken part in founding it in the fall of 1923 and was later appointed honorary president of the society. Einstein also played an active role in the Russian Natural History Week in Berlin, which took place in the halls of Berlin University in June 1927. Several prominent Soviet scientists participated and gave lectures; among them the physicists A. F. Joffe and P. P. Lasarev.

FH: During your time, Einstein undertook several major trips abroad. First of all, he traveled to Pasadena as guest lecturer at Caltech, but, apart from that, he also made many trips to lecture at congresses, particularly in Belgium, Holland, and England. The wintertime journey to Pasadena was repeated twice.

HW: The first time, I could have come along, and even ought to have come so that I could help Frau Professor in the household. But I just could not make up my mind. Today, I regret it deeply, because later on I never had another opportunity for an extensive journey.

FH: And I also regret it, for if you had gone you could now tell us about many interesting experiences with the Einsteins. Fräulein Dukas traveled with them, in your place.

HW: As Professor Einstein's secretary, she was surely more important. But I had another reason why I was not thrilled at the suggestion; I did not speak a word of English.

FH: That was not a good reason for declining. Einstein also knew very little English at that time, even though he had tried to learn that language since 1913. Later, even after Einstein had lived in Princeton for several years, his collaborator Leopold Infeld estimated Einstein's vocabulary of English words at no more than three hundred. When he talked to colleagues who did not know German, Einstein communicated, as Infeld put it, "in a language that he believed to be English." In the biography *Albert Einstein—Creator and Rebel*, Helen Dukas describes Einstein's spoken English as "droll," and he fared even worse with written English. During the Second World War, when Max Born wanted to correspond with his friend in English, Einstein answered, "I cannot write in English on account of its devious orthography. When I read [a word], I hear it in my mind's eye, and I do not remember what the written word looks like." The published facsimiles of texts Einstein wrote in English reveal his great insecurity in English orthography. Apparently, however, he also had difficulties with French, though he felt more at home in it. In 1915, following a meeting with Einstein, Romain Rolland wrote this comment in his diary: "He speaks French with difficulty and mixes German expressions into it." It seems that Plesch is right when he observes that Einstein had "no talent for languages."

HW: I had the impression that Herr Professor had trouble with foreign languages, for when foreigners came to visit Frau Professor conducted most of the conversation and also acted as interpreter. She was evidently better versed in foreign languages than he.

FH: If I remember correctly, you mentioned to me at one time that Einstein was interested in Esperanto.

HW: It was something that meant a lot to him, for this language was to serve as the means of communication for all people. There were many discussions about it in the Haberlandstrasse, and that is why I know something about Esperanto.

The issue of an artificial language that could serve as an international auxiliary language was widely discussed in those years, although not as vigorously as before the First World War. At that time, the chemist and Nobel Prize winner Wilhelm Ostwald was a very energetic promoter of an improved Esperanto, called Ido, and urged its use in the scientific literature. At least, the abstracts of articles in specialized journals were to be translated into the artificial language. However, he had no success with this proposal.

FH: Was Einstein himself an Esperantist? Did he learn it or speak it?

HW: That I cannot say, but I do know that Herr Professor took great interest in it.

FH: Due to our concerns with Einstein's linguistic difficulties and his interest in Esperanto, we quite lost sight of the first journey to Pasadena. Was there a lot of excitement in the house, maybe travel nerves?

HW: The first time, yes, but later on, not any more. Herr Professor took no part in the travel preparations, however; Frau Professor took care of packing the suitcases, and I helped her with it. For longer voyages, the Einsteins used steamer trunks in which dresses and suits could be hung on hangers. Upon their return from Pasadena in March 1931, these trunks were much fuller than on their departure. That was the first time I saw Herr Professor dressed up to the hilt. In the photos that were shown around afterward, one saw what an elegant impression he made when he was dressed appropriately. I heard that before their departure from America several crates of oranges and a whole sack of raw coffee beans were brought aboard for Herr Professor. In Caputh, we from time to time took the raw beans to Potsdam for roasting—about twenty pounds at a time. Frau Professor had organized that. The weight is naturally reduced in the roasting, but a little over ten pounds remained each time. Whenever we roasted, I always got two pounds, and a packet was sent to my mother, as well. Herr Professor, who otherwise made only decaffeinated coffee for himself, also drank an occasional cup of bean coffee now, in the afternoon.

FH: The Einsteins probably brought you a nice memento of their journey.

HW: Yes, a purse.

FH: Finally, for today, another question about the Haberlandstrasse apartment. Supposedly, there was a bowl with signed photos of Einstein on the piano in the living room, that were intended for visitors who wanted his picture and autograph. Was that really the case?

HW: Not while I was there. It might have been earlier. I only know that on the piano was a silken, fringed cover and on it stood a large vase, which was not filled with flowers, however.

FH: An offering of such signed photographs might have suited Frau Elsa's personality, but certainly not Einstein's, since he detested the "personality cult," as he called it. He made an ironic poem about people who wanted a photo with his scribble on it. It goes like this:

> Wherever I go, wherever I stay,
> A picture of me is ne'er far away,
> On the desk, and on the wall,
> As a pendant—I've seen it all,
> Everyone, it makes me laugh,
> Must go and get my autograph,
> Owning his scribble gives them joy
> For he is such a learned boy.
> My good fortune has reached such height,
> that I ponder, when the time is right:
> Are you, yourself, perchance insane,
> Or are the others all inane?

Wo ich geh und wo ich steh,
Stets ein Bild von mir ich seh,
Auf dem Schreibtisch, an der Wand,
Um den Hals an schwarzem Band,
Männlein, Weiblein, wundersam,
Holen sich ein Autogram,
Jeder muss ein Kritzel haben
Von dem hochgelehrten Knaben.
Mensch, so frag in all dem Glück
Ich im rechten Augenblick,
Bist verrückt du etwa selber,
Oder sind die andern Kälber?

HW: I already know this poem from one of your books.

FH: I quoted it in my 1963 Einstein biography,[39] and it comes from a celebratory volume in honor of Einstein's fiftieth birthday that was published privately in Berlin, in March 1929. It appeared as a one-time edition of eight hundred copies and is entitled *Occasional Pieces by Albert Einstein.*[40] The vignette on the cover—Einstein's head in profile—was drawn by Margot's teacher Kurt Harald Isenstein, who also made a bust of Einstein. Here is that little volume. I acquired it many years ago in an antiquarian book shop for two or three Marks, but today it is a bibliophile's treasure. It contains very remarkable observations by Einstein, some about the psychology of research, some about ideological issues. They reveal the astonishing breadth of his vision, but they also show the limits of his political awareness. But the thesis formulated by Einstein regarding the antiscientific atmosphere in fascist Italy remains unconditionally valid: "With dictatorship comes the muzzle, which leads to stupor. Science can only flourish in an environment of free speech."

FOURTH CONVERSATION:
SUMMER HOUSE AND SAILBOAT IN CAPUTH

FH: You told me that the Einsteins lived in Caputh in the spring and summer of 1929, even before their timber house was inhabitable.

HW: We stayed for the whole summer of 1929 in the Potsdamer Strasse, while their house was being built. It was an old house with a large garden, and it was also old in that it did not have city gas, and I had to do all the cooking on a coal stove with only two small burners. The house was directly on the shore, on the right side when coming from Potsdam. Frau Professor often walked up to the building site on the wooded slope, to check on the progress of the work. I often went along with her—indeed, quite often.

FH: On your visits to the site did you get to meet the architect?

HW: The only thing I remember is that Frau Professor talked with the site manager of the firm from Niesky.

FH: The house was apparently constructed with great dispatch.

HW: Yes, the construction came along very quickly. I think that we moved into the upper floor by the fall of 1929. The unseasoned wood used to panel the walls gave a wonderful smell to everything; that is something I remember very well.

FH: Antonina Vallentin confirms your impression. She writes from personal knowledge that the house was finished in October 1929 and that "the walls of the spacious living room were paneled with thin light-colored slats, the kind used for jalousies." Later on: "The new wood made the house smell like a saw mill."

HW: Yes, that's the way I remember it.

FH: A few questions about the prehistory of the house: You know my article "Albert Einstein and the Political Fortunes of His Summer House in Caputh, near Potsdam," which first appeared in a history-of-science journal and, later, in extended form in my anthology *Einstein and His World View*.[41] It presents the history of the house based on archival material and on oral accounts. I would now like to hear from you what you recall of its history. Did the family talk about the City of Berlin's wish to present to the professor for his fiftieth birthday either a house on the water or a vacant building lot? Did you hear about the series of suggestions made by the municipal authority, none of which could be realized because the city did not own the properties outright? It was, finally, suggested to Einstein that he should select a suitable building lot, which the city would then buy for him. Frau Elsa found the lot in Caputh. It belonged to a builder, who had his own summer house next door. Did you hear about all that?

HW: Not in such detail. But I know that they discussed the newspaper accounts of the intended birthday present of the City of Berlin. There was quite a lot of excitement. Herr Professor was very annoyed that it all dragged on so slowly and that new obstacles kept appearing. There was a lot of grumbling about it in the Haberlandstrasse, as well as later in Caputh, when we lived in the Potsdamer Strasse. That I recall very well, but at the time I was unaware of the whole prehistory, which is quite complicated, according to your account.

FH: The Einsteins bought from the forestry administration a strip of land that gave them access to the carriage road and added it to the plot the municipality wished to give them.[42] This is the plot on which the house was then erected.

HW: I had not heard about the additional land purchase at the time, but I have a dim memory of them buying the narrow strip that connects the lot to the Waldstrasse. At the very beginning, one could reach the plot only from above, from the carriage way, and the house could not be reached from the Waldstrasse. The sloping ground on both sides of the stairs leading down from the carriage way to the front door was planted with small pines to maintain its woodland character.

FH: Since you witnessed the construction of the timber house from the ground up, and lived in it during several summers, you can probably describe how the rooms were arranged.

HW: After you came down the steps from the carriage way, you had to ascend a few steps to reach the house entry. When you entered you were in a fairly long corridor running sideways, its floor covered with light and dark tiles, arranged as on a chess board. The kitchen was to the right and straight ahead was the large living room. On the left, underneath the stairs leading to the upper floor, were steps that led down to the cellar, where the furnace for the central heating was located. It was fired with coke, as was still customary then, and I had to feed it continually, even at the height of summer, so that there would always be hot water. Herr Professor took a hot bath every morning, just as in Berlin, as did Frau Professor. Hot water was also needed in the kitchen for doing the dishes. The hot water tank was started up each morning by a man from Caputh, and I just had to add more coke so the water would stay hot. I can still recall the big furnace very clearly because I poured coke into it all the time.

FH: According to the plans and elevations that its architect, Konrad Wachsmann, published in his 1930 book *Wooden House Construction*, the house had six rooms, an open living space, and two terraces, the lower one roofed-over and the upper open to the air. In addition, there was the kitchen, a bathroom, etc. Could I ask you to describe the rooms and their furnishings?

HW: Following the corridor to the left, Herr Professor's room was at the very end. It was both his study and bedroom. To the left, as one entered, stood his bed—a real bed, not a couch or a divan—inside an alcove that could be closed off by a curtain. At the window was a simple flat desk with one drawer, and to the right a broad shelf with books and reprints. The violin was also usually on that shelf. Ahead of Herr Professor's room was the bathroom—it had a round window—and then came Frau Professor's room. Here, too, the bed stood in an alcove, and there was also a small ladies' desk. Then came the living room that I mentioned before. Its three French doors were usually open and led out to the roofed terrace. It was a lovely, spacious terrace where one could sit in comfort, for there were always chairs and deck chairs on the terrace. From there, three or four steps led down to the garden, which was terraced like a rock garden, with a paved walkway running down the middle. When we came out in the spring, rock cress and forget-me-nots carpeted the garden in blue-and-white, and I can still see that picture before me today—it was always very lovely. There were also daffodils among red primroses and, quite generally, lots of spring flowers. I recall the flowers in the garden and lower terrace especially well because I took care of them. Fuchsias were always planted in the open living space underneath the living room (i.e., in the shade), and in the upper terrace, which was sunny, a mix of blue lobelias and white petunias. It was a magnificent spread of blue and white.

FH: Was the summer house occupied that early in the spring?

HW: Right after Herr and Frau Professor returned from Pasadena, at the end of March, we generally were no longer in the city apartment. It was then time to be off to Caputh, and we often stayed until the beginning of November—or even longer—and were still there when the leaves turned a bright yellow. The move back to Berlin was always delayed as long as possible.

FH: Rudolf Kayser wrote in his 1930 Einstein biography that Einstein took such pleasure in his summer house that he, at first, considered moving his residence to Caputh. Do you think this possible?

HW: Absolutely. Herr Professor was indeed genuinely enthusiastic about the house, which was situated so delightfully on the wooded slope with a panoramic view of the Havel lakes.

FH: In the end, Einstein did not carry out that plan for various reasons, mostly because of his commitments in Berlin; but please, continue your description of the rooms, starting with the large living room.

HW: As you entered the living room from the corridor, there was a fireplace on the left, with a longish, rectangular table in front of it, along with a few chairs. In the center of the room stood a large round table, made of bare wood with a white top surface, together with chairs. Everything fitted in well with the rural decor. At mealtimes, I covered it with a table cloth.

FH: What else was in the living room?

HW: I recall three or four small nesting tables to the left of the French doors. These could be taken apart, in case one wished to put something down. To the right of the terrace door was a small table for books and magazines. The floor was varnished—not painted—and in the center of the living room was a rectangular raffia or straw carpet or, possibly, coconut matting. That was the only carpet I recall in the summer house.

FH: Antonina Vallentin recalls seeing an enormous Japanese wood sculpture in that room, supposedly consisting of tree limbs and branches and inhabited by animals and dwarfs, an artistic creation with the delicacy of ivory carvings. It was, supposedly, a present from an exotic admirer of Einstein's. According to Antonina, Einstein had no space for it in the Haberlandstrasse, but in Caputh it was even more out of place. Can you recall a monstrosity of that kind? With dwarfs and animals on tree branches?

HW: I really cannot recall seeing anything like that.

FH: Let us assume that some kind of mix-up took place. But could you please say something about the upstairs rooms.

WH: When you came up the stairs to the upper floor, you faced the guest room which was usually occupied by the Kayser couple. Occasionally, other visitors would also stay there, such as Professor Ehrenhaft with his spouse, or the elder son, Hans Albert, with his wife. There were two beds in the room and adjoining it was a bathroom with a sit-down bathtub. Next to it was a slightly smaller room, which Margot occupied when she was in Caputh. Her husband was rarely out there. This is also where the seamstress sometimes did her work. Next to it was a small room, which was my little domicile. It had a window that looked out on the upper terrace, which was considered the sun terrace and its floor was wooden decking. You could enter it through a door next to my room, but you could also reach it from below, using the outside stairs with their white railing. Wooden railings, painted white, also ran around the upper terrace. Ilse and Margot, in particular, used to spend a lot of time lying on the terrace. The deck chairs were brought out for sunbathing; they were not there all the time.

FH: The property presumably had its own supply of drinking water, as is customary for larger properties today. Or was city water already up there, at the edge of the forest?

HW: There was proper city water, as well as city gas and electricity. I had a gas stove in the kitchen, just as in the city apartment.

FH: Can you say something about the light fixtures. Were there any colored lights or a forged iron lantern outside on the terrace?

HW: It was not like weekend houses nowadays. If I remember rightly, there were globular light fixtures on the terrace, and the rooms had mostly ceiling lights. Herr Professor's desk was next to the window, which reached all the way to the floor, and on it was a simple desk lamp with a flat porcelain shade, as was customary then.

FH: Did Einstein spend much time working in his room?

HW: It was quite irregular, as in Berlin. Sometimes, he would stay in his room to work for hours. Sometimes, he got up very early and headed for the water, or he went walking in the woods, usually by himself. In the summer and in the fall he always brought lots of mushrooms home from the surrounding woods, mostly boletus and morels.

FH: Did Fräulein Dukas come to Caputh regularly to do the correspondence?

HW: Fairly regularly, but not every day.

FH: Where did the guests and visitors sit?

HW: In the living room and, if the weather was nice, on the big terrace. When his closest associates came to visit, mostly colleagues like Geheimrat Planck or Professor von Laue, Herr Professor also took them to his room. The living room was mostly used at mealtimes. Herr Professor would take particular guests on walks to show them the lovely countryside, or he took them sailing on the Templiner Lake.

FH: At this point we should say something about the sailboat that was presented to Einstein for his fiftieth birthday by wealthy Berlin friends, by bankers. According to some biographies, the boat was delivered when the timber house on the wooded hillside was already occupied.

HW: That may well have been the case. I distinctly recall walking from the house down to the dock to see the boat. The builder of the boat was a man called Harms, and he had delivered it personally. I assume that this photo comes from him.

FH: The photo shows Einstein as the proud owner in front of his boat. The back of the picture reads, "In remembrance of *Tümmler*'s builder, Adolf Harms." I found out from archival records that the boat was built especially for Einstein, in a boatyard in Friedrichshagen on the Müggelsee. It was not part of a serial construction.

HW: That is something I did not know. But I know that it was called *Tümmler*. Yes that was the *Tümmler* [*Porpoise*].

FH: This boat played a big role in Einstein's life in Caputh. Later, in his emigration, he missed it sorely. Could you describe it?

HW: Yes, of course, I got to know it very well, for I always used to clean it. It was beautifully finished inside. When you entered the cabin from the deck, there was on the one side a little cupboard with plates, cups, and cutlery for two or three persons, also a cooking stove. On the other side was the toilet. Then you came to the two sleeping benches, separated by a central gangway. As I recall, they were covered in artificial leather. You could sleep on board, for there were beds and blankets, but Herr Professor did not overnight on board. There was also a built-in auxiliary engine in the stern.

FH: According to the photo, the boat had three oval portholes on its sides.

HW: It was a beautiful, comfortably furnished boat.

FH: When the boat was confiscated in August 1933, the official police report described it as a cabin yacht, constructed of solid mahogany. Its length was seven meters (23 ft.), its beam about two-and-a-half meters (8 ft.), with a sail area of twenty square meters (210 sq. ft.). It was evidently quite a large boat. In his letter to you on April 1947, Einstein calls it "the plump sailboat."

HW: The boat was quite broad, and it sat securely in the water. You could stand up in the cabin and move about comfortably.

FH: Did Einstein always return before dark?

HW: If he was becalmed, Herr Professor waited among the reeds, sometimes for hours, until he could sail on. He rarely used the built-in auxiliary motor. He would sail into the reeds and would sometimes lose track of the time. It's simply that he was often very dreamy and lost in thought. Now and then it happened that he returned home very late, and we were all quite worried. I remember one occasion when Frau Professor and I walked all the way down to the steamship landing in Caputh and looked out for his boat for a long time. Somebody went out in a small fishing dinghy to look for him. Such excitements happened quite often. Herr Professor sailed mainly on the Templiner Lake, less on the Schwielow Lake.

FH: Did you sometimes go sailing with him? Did he sometimes take you across the lake to Potsdam on your free afternoon?

HW: Herr Professor invited me repeatedly to sail with him when I had time, because I always cleaned the boat, but I never did. Boats did not agree with me, and, besides, I got to Potsdam much more quickly by bus than on the sailboat.

FH: Several biographies are of the opinion that Einstein did not know how to swim and that may be true. When he was young, it was not as common to teach swimming, as it is today. But did he go bathing, at least? After all, there was a swimming beach in Caputh and there was surely a bathing place nearby.

HW: Herr Professor only went sailing, often with a bare torso and in his blue training pants—actually dark blue—and wearing only sandals, without socks. Sometimes he went down to the boat completely barefoot. They used to joke in Caputh that only Professor Einstein went barefoot there.

FH: Did you go bathing? Did Frau Elsa and did the two daughters bathe in the Templiner Lake?

HW: In good weather I went bathing every evening. That was when the water was warmest. But I always went by myself or with my cousin when she came from Berlin. The Einsteins never came along. I don't think that any of them knew how to swim, and I never saw a swimsuit hung up to dry. It could be, of course, that they bathed in the Baltic Sea when they stayed on Hiddensee. That I could not say; but in Caputh, certainly not.

FH: There are several accounts by scientists who had gone sailing with Einstein on the Havel Lakes. Max von Laue, who often sailed with Einstein, remarked that Einstein paid little attention to sportive correctness when he sailed, and it did not matter to him if his sails luffed. The physicist Erwin Schrödinger, who succeeded Max Planck at the University of Berlin, also often visited Caputh with his wife and went sailing with Einstein. Annemarie Schrödinger wrote to me in 1964, a year before she died: "We were often guests of Einstein, in his Berlin apartment, as well as in Caputh. The visits to Caputh were particularly enjoyable for my husband, because we would go sailing with Einstein for hours, when the two men could have scientific conversations without being disturbed."

HW: I cannot recall the name Schrödinger.

FH: That could well be. Schrödinger had moved to Berlin only recently and was not nearly as popular as Planck or von Laue, who had been at the University for a long time, Planck since 1889.

HW: Those two names I recall very well.

FH: The physicist Philipp Frank from Prague, whose Einstein biography I have cited several times, was also among Einstein's guests in Caputh. In his book he described the wooden house and the lovely view from the living room through the glass doors. But you would hardly be able to recall this rare guest.

HW: Professor Frank's name does not ring a bell. Unfortunately. It all happened so long ago, after all.

FH: That reminds me of something that happened around 1930. A renowned experimental physicist who also taught in Berlin, and also was a sailing enthusiast, often sailed on the Havel Lakes with his little son. Once, when he came close to Einstein's boat, he said to his boy, "Look, over there, that's Uncle Einstein." The boy, at the inquisitive age, looked at the sailboat for a while and asked, "Daddy, why is Uncle Einstein, in fact, an aunt?" This childish question was surely justified in those days when men usually wore their hair short; but it would hardly be asked today.

But now to something else. When I began work on my Einstein biography in the nineteen fifties—it appeared in 1963 as *Albert Einstein—A Life for Truth, Humanity and Peace*—I asked some of the older inhabitants of Caputh what they remembered of Einstein.[43] They told me, among other things, that he fairly regularly went sailing with a blond lady, evidently his lady friend who often came to Caputh.

HW: She [Margarete Lebach] was an Austrian, who, as far as I know, had only recently moved to Berlin. As I recall, she came to the Haberlandstrasse for the first time around 1931 and brought wonderful vanilla crescents (*Vanillekipferln*) that she had baked personally. They were wonderful specialties, a genuine Viennese pastry. She brought them for Frau Professor, who savored this kind of thing. I suppose that a closer friendship developed between her and Herr Professor. In any case, during the summer she came to Caputh once a week, and, whenever she came, Frau Professor would always travel to Berlin early in the morning to shop and to take care of other errands. She would always leave for the city early in the morning and did not return until late in the evening. She relinquished the field, so to speak. The Austrian was younger than Frau Professor, and she was good-looking and cheerful, loved to laugh a lot, as did Herr Professor, and she spoke with a Viennese dialect. Once there was a lively discussion about her in Caputh, in the absence of Herr Professor. I overheard it and could not help hearing it because it was carried on rather loudly in the living room, which was next to the kitchen, after all. The daughters told their mother, "You simply have to put up with it, or you must separate from Albert."

FH: It had come to that?

HW: Yes. On that occasion they only spoke of Albert, not as usual of *Vater* Albert. The mother had remonstrated about it to her daughters, and later I heard Frau Professor crying. But she came to terms with it and traveled to Berlin on the specific days to avoid an encounter. But after that, the close friendship between the two women was over.

FH: That one can imagine. How long did this relationship last?

HW: It could be that it lasted only one summer, the last summer in Caputh.

FH: She sailed for only one summer, one might say—alluding to the title of the well-known film.[44] The local population seems to have paid a good deal of attention to the habits and relationships of the famous scientist.

HW: I also heard a good deal about this only from the people in the village, where it was an open secret. One could plainly see, after all, who was on board when the boat sailed out. That could not remain hidden. As I said before, Herr Professor just had a weakness for beautiful women.

FH: There is also the fact that his woman friend came from Austria. Einstein found the Austrian temperament particularly congenial, as is also borne out by his friendship with Ehrenfest and Ehrenhaft. That empathy may also have played a role. But the deeper reason for Einstein's conduct is revealed, in my opinion, in the draft of a letter I recently came across in an auction catalogue of a West German autograph dealer. It is Einstein's response to a lady who was distressed by her husband's infidelity and had asked Einstein for advice. He consoled her with these words: "You probably know that most men (and not a few women) are not naturally disposed toward monogamy. Nature breaks through all the more forcefully when custom and circumstances place obstacles in that individual's path.

Enforced faithfulness is a bitter fruit, however, for all concerned." Einstein, who was normally an outsider, evidently did not differentiate himself from the majority of men on this issue, nor did he place any constraints on his own nature—in this, as in everything else.

HW: That may well be.

FH: Besides, Frau Elsa was no longer in her flowering prime. This is how Antonina described her from personal observation: "Her face has become squishy, her hair has turned prematurely grey. She, who had long chided her husband for neglecting his appearance, has now let herself go in the same way. While putting on her hat, she whisked away, like flies, the heavy skeins of grey hair that fell in her face." Judging from contemporary photos of Frau Elsa, I consider this description apt.

HW: Frau Professor was in my opinion a pretty woman, even at that time, and she was also deeply attached to Herr Professor.

FH: Her vanity and craving for recognition did get seriously on his nerves. Some remarks by Einstein to his Soviet colleague Joffe leave no doubt about it. In his memoir, *Encounters with Physicists*, which I mentioned before, Joffe writes of Frau Elsa, that she had not become a close friend to Einstein and that she used every means to undermine Einstein's efforts to stay clear of all manifestations of his worldwide fame. Joffe adds: "I have decided to write about this only because Einstein himself told it to me very explicitly." This conversation probably took place at the end of the Twenties—that is, in your time—when Joffe often stayed in Berlin.

HW: But Frau Professor also had her good points; that is something I want to stress again and again.

FH: Let us now turn to some practical matters concerning Caputh. How was the shopping handled? Did you go into the village? Or travel to Potsdam to buy provisions?

HW: Almost everything was delivered by a firm in Potsdam. We placed the orders over the telephone in the office of the master potter [*Töpfermeister*] Wolff, and usually I did this. Whatever was ordered was then delivered, including the provisions. I rarely had to go shopping in Caputh; everything would have been much too far for that. The baker came and delivered fresh rolls, driving from one home to the next. Everything was delivered to the house, just as in the city apartment, and I did not have to bother.

FH: What about the Caputh menus?

HW: They were all simple dishes. When there were casual visitors, we generally had green beans and matjes herring filets at the midday meal. That was served often. New potatoes had their skins peeled before being tossed in butter and parsley. I remember that very distinctly, and I especially want to mention asparagus again, which was available in any amount in Caputh.

FH: You said that you placed orders over the telephone in the master potter's office. Was there no telephone in the Einstein house?

HW: No. If we needed to place a call, we went down to the master potter's office, and when a call came in we were called to the telephone. At first, they simply shouted up to the terrace, but eventually we had a signaling system that used a small trumpet that Frau Professor bought specially for that purpose. Three short trumpet blasts meant that I could take the call, two longer blasts meant Frau Professor had to come down and provide some information, and one long, loud blast meant that Herr Professor himself was requested on the telephone. As I said, that was only toward the end. I recall going down to the office many times and telephoning Potsdam to order provisions. Sometimes I also had to call Berlin to deliver a message. There was always something that needed to be ordered or arranged.

FH: I assume that Einstein did not have a telephone installed on purpose? He could surely have obtained a connection without trouble.

HW: If possible, Herr Professor just wanted his peace and quiet. I cannot recall that he did much telephoning in Caputh or in the city apartment. It had to be something quite extraordinary. Otherwise, Frau Professor took care of everything.

FH: Did Einstein do any gardening on his own in Caputh or do any physical work on his property to make up for the cerebral work at the desk?

HW: Absolutely not. At least, I never saw Herr Professor planting anything or digging. That was all taken care of by the gardener.

FH: That would contradict the account of Einstein in Caputh that the

Soviet theater director Natalia Saz gave in her book *Novellas of My Life*. She describes Einstein as a Sunday gardener who busied himself with garden hose and clippers, surrounded by shrubs and beds of strawberries, cucumbers, potatoes, and cabbages.

HW: This account is erroneous. There were no beds of strawberries, cucumbers, potatoes, and cabbages on the property, no agricultural cultivation at all. The strawberries were bought from the neighbor or down in the village, and so were potatoes and cucumbers. I recall that in the very beginning we did have a small bed of kitchen herbs; that I remember distinctly. But that is all we had.

FH: Natalia Saz also reports that there was an "old piano" in the large living room and that she accompanied Einstein's violin playing on it. Was there such a piano?

HW: As I remember it, there was no piano in the large living room, or anywhere else in the summer house, and certainly not an old piano. I don't think I am mistaken in this.

FH: Einstein's former neighbors expressed the same opinion when I interviewed them. They had often heard Professor Einstein playing the violin but couldn't recall the sound of a piano. Did Einstein really play the violin regularly, at a particular time of day, on the terrace, as one of our magazines reported some time ago?

HW: I read that, too, and all I can say is: No, that is utter nonsense; that is not true. Everything Herr Professor did was always quite spontaneous. He played the violin at no particular time of day, neither in the city apartment, nor in Caputh.

FH: The quality of Einstein's violin playing is controversial and has been judged unevenly. We mentioned already that, as one book had it, Einstein's playing was pathetic. That may be a tactless statement, but it reminds me of the opinion expressed by Professor Walter Friedrich, the former rector of our University and later president of our Academy of Sciences. When I asked him in 1964 what he remembers of Einstein, he said literally, "He bowed like a lumberjack." Since Friedrich was not only a notable radiation researcher but an excellent violinist, his judgment carries some weight.[45] It is strengthened by János Plesch, who wrote of Einstein:

"He never achieved virtuosity and many players are more talented as technicians than he." Plesch faulted Einstein for his bow control, but he also reports the following incident: Einstein once fantasized for hours in Gatow on his (Plesch's) organ, which was housed in a pavilion close to the shore. Without Einstein noticing it, a rapt audience in dinghies and yachts had assembled on the Havel to listen to these wonderful harmonies, without any of the listeners knowing or suspecting that it was Einstein who was playing the organ.

Did the Professor play the violin at midnight in the rooms and hallways in Caputh, as he sometimes did in the city apartment?

HW: No, I would surely have heard it.

FH: Can you confirm Plesch's claim that Einstein possessed a considerable collection of violins that admirers had presented to him over the years? Can you recall such a collection?

HW: I only know of a single one, and Herr Professor was very fond of it.

FH: Our handsome János exaggerated a little in this, and not for the first time; for if one were to believe his expositions, there were no prominent scientists, politicians, painters, sculptors, conductors, and actors with whom he was not, more or less, befriended. He was indeed "an overcommitted celebrity physician whose need for recognition was inexhaustible," which is how he was characterized in an appraisal I discovered in the archives of Humboldt University. When the Swiss Einstein biographer Carl Seelig describes Plesch's book as being "replete with fantasies," there is some truth in it. On the other hand, here and there, János can be believed, as we have indeed done in our conversations. But let us return to Caputh. Did Einstein have close relations with residents of the village? Can you recall any citizens with whom he liked to chat?

HW: I would name, first of all, the typesetter Meier on the Potsdamer Strasse. He was a Social Democrat and, during the depression, he was out of work. Herr Professor often chatted with him and also gave him work to do in the garden, to help him financially. Quite generally, Herr Professor conversed a good deal with the inhabitants of the village. He was very genial [*leutselig*], as one says, and always very considerate toward others.

FH: Toward you, as well, I suppose.

HW: Yes. When he walked up from the bus stop with a suitcase, and I ran toward him to help carry it, he turned me down with the comment, "I surely won't let a woman carry the suitcase for me." When I did help carry the suitcase, I always had to pretend that I only held on to the handle with him. When he walked down to the bus stop, it was the same. But usually he traveled to Berlin by car.

FH: Now, with regard to the incident in which Frau Elsa's lorgnette played a certain role: You told me how, on that occasion, you walked to Caputh with the Einsteins in order to vote in a general election [*Reichstagswahl*], of which there were many in the last few years of the Weimar Republic—in 1932, there were even two. Please describe the incident.

HW: The polling place was a tavern in the village, as was customary. Already on the way down, Frau Professor asked me how I was going to vote. I said SPD [Social Democratic Party], and she told me that she would do likewise. Several people in front of the polling place displayed posters or carried signs that exhorted people to vote for this or that party. Frau Professor stood in front of the SPD poster, carefully checked it out with her lorgnette that she wore on a chain, and then said quite loudly, "Yes, that's how we will vote."

FH: Did Einstein also come to vote?

HW: I have tried to recall if Herr Professor had come along; it could be that he was away on a trip. It almost seems that we all went together to vote, but I cannot be sure today. But I am quite certain that Frau Professor pointed at the SPD poster and said to me, "That's how we will vote," speaking so loudly that all bystanders must have heard her.

FH: Now for something that has nothing to do with general elections or election posters. You mentioned in our conversation that a certain long-haired dachshund had been much loved in Caputh.

HW: Purzel actually belonged to the master craftsman [Herr Teichmann] who lived down below in the Waldstrasse, but he was always welcome in our home and liked to accompany us wherever we went. One time, when Herr Professor was being driven to the observatories in Potsdam, Purzel jumped into the car without anyone noticing it. He often accompanied Herr Professor on his walks in the woods or he would lie at his feet on the terrace. He simply was part of the household and the Teichmanns didn't see much of their dog during the summer. Herr Professor was always very glad whenever Purzel joined him on a walk.

FH: You also told me of a tomcat that suddenly appeared and then did not budge.

HW: He appeared out of nowhere and stayed. He was white with gray and black blotches. I took care of him and, to please me, he was allowed to remain in the house. He was still young. When he walked across the terrace and Herr Professor happened to sit there, he would call him and pet him—yes, that was Peter.

FH: Where was Peter's basket?

HW: He always spent the night with me. He jumped into my room from the upper terrace through the window—that was a piece of cake for him—and then lay down in bed with me. There were times when that little chap had me shedding bitter tears, when he fought with other tomcats and was bitten. I would then get up in the middle of the night and make chamomile tea to wash out his bleeding ears. That reminds me of something: We had bought kippers in Potsdam and Frau Professor put a whole kipper on the floor for him. This really frightened the tomcat. He circled the kipper for a long time and finally jumped and bit the kipper in the neck, as if it were still alive. That sticks in my memory. We were all very amused, and Herr Professor laughed about the tomcat that was afraid of the big kipper.

FH: When did he come to you?

HW: In the spring 1932, when he was still quite young, but by autumn he was already fighting strenuously with the other tomcats. Later, the woman next door was paid a boarding fee for feeding him during the winter.

FH: Did the tomcat get along with the dog, with Purzel?

HW: They both got along very well and quickly got used to each other. Neither of them was jealous of the other, as sometimes happens with house pets.

FH: Was the blue parakeet brought from the city apartment to Caputh for the summer?

HW: No. I think the Kaysers always took care of it whenever Margot and her husband were not in Berlin, because one could hardly leave the little chap alone.

FH: Did people get up later in Caputh than in the Haberlandstrasse?

HW: That was quite variable and there was no particular time. We ate the midday meal between twelve and one, when everyone was at home. Here, too, I ate in the kitchen. Right afterward, Herr Professor usually went sailing again or went for a walk in the woods.

FH: Another question regarding the timber house: What was the color of the roof?

HW: I recall it as being red. The red color went well with the dark-stained wood of the house, though the grain of the wood was still discernible. The window frames were lacquered white.

FH: Did the Einstein celebrate the topping off or the initial move into the house in any way?

HW: Oh dear, no, not at all. Quite generally, there wasn't much celebrating at the Einsteins. But I do want to mention a small occurrence that my cousin recently reminded me of. Once, when she came out to Caputh, Herr Professor happened to be sitting on the terrace in his bathrobe. As he got up to shake her hand, the bath gown opened up—and he was wearing nothing else. My cousin turned beet-red with embarrassment, whereupon Herr Professor asked her, "How long have you been married?" She answered, "Ten years." He asks again, "How many children do you have?" She says, "Three."—"And you still blush?"

FH: Einstein was just not puritanical; that is known from several reports. But didn't you blush, as well, when in the morning in the city apartment you occasionally saw him walk from the bathroom to his bedroom in the buff? You had once alluded to that.

HW: Whether I blushed, I can no longer recall, but it was very embarrassing for me. Either Herr Professor had not bothered putting on his bathrobe, or he was too lost in thought to remember to wear it.

FH: I suppose your cousin came quite often to visit you in Caputh?

HW: During the summer, she came almost every Sunday. Frau Professor was glad of it because it meant that I would not be going to Berlin. Sometimes, she also brought one of her children with her. I can almost see it before me, how Herr Professor, on the lower terrace, let the little girl—she was about four then—listen to the ticking of his wrist-watch. We usually sat on the sun terrace. Frau Professor was anxious that I, too, should be able enjoy the lovely sojourn in the country.

FH: Did your mother also come to Caputh?

HW: Yes. I told you already, how once Herr Professor, on the terrace, saw my mother walking up the path, and how he called to me in the kitchen, "Fräulein Herta, mother is coming!" He ran down immediately to meet her and to carry her bag. She had a doctor's bag, as they were called, and he did not want to see a woman carrying the bag. My mother came to Caputh quite often, but she did not spend the night there.

FH: You mentioned that Frau Elsa wanted you to enjoy your country sojourn. Was your work load in Caputh not as great as in Berlin?

HW: I had a great deal more free time in Caputh.

FH: You always took the bus when you wanted to go to Berlin.

HW: By bus to Potsdam and then, with the S-Bahn, by way of Wannsee, to Berlin. Herr Professor also took the bus to Potsdam and then the S-Bahn when a car was not at his disposal. That reminds me of something: Once, I arrived at the Potsdam main station from Berlin after the last bus had already left. I had no choice but to walk to Caputh in the middle of the night, along the shore of the Templiner Lake. It must have taken me between one-and-a-half and two hours. It was pitch black. Suddenly, in the middle of the path, a small critter jumped up. I was so shocked that I broke into a sweat. I had just been at the hair dresser in Berlin, where I had my hair curled, and in the morning all the curls were gone. When I told this to Herr Professor, he said that he had heard of a person's hair turning gray from a fright but never that one's hair could lose its curls, that was news to him.

FH: Did Einstein's physicians visit often in Caputh? For instance, the house doctor.

HE: I cannot recall a visit of Sanitätsrat Dr. Juliusburger. Professor

Plesch also came only rarely to Caputh, but Professor Katzenstein came fairly regularly, for there was a genuine friendship between them.

FH: Another question that has to do with the political developments at that time. Did you notice anti-Semitic tendencies in the last year in Caputh? The Nazis were, after all, very intolerant toward Jews even before the dictatorship was created, and everyone knew that Einstein was a Jew.

HW: Herr Professor was very popular among the inhabitants of Caputh. I don't believe that there were any exceptions.

FH: The reason I ask is that Antonina Vallentin describes an incident in her Einstein biography that supposedly took place during her last visit to Caputh in May 1932, an incident in which you played a role. I will read you that section:

> The housekeeper entered and remained as if she had something to say. She had been with them for a long time and Elsa encouraged her to go ahead and speak. She said: "I will never go to our baker again. He made a very ugly comment to me." She hesitated: "Yes, he said that he cannot understand how I can continue to stay with Jews. That sonovabitch!" [*Schweinehund!*] She trembled with rage. In response to Elsa's question, she said: "Oh no, Herr Professor is not at risk, they are all much too fond of him here. It is only that sonovabitch!" Her face was convulsed with shame. I said: "But there might be others, as well." She gave me a distraught and suspicious look and silently left the room.

Can you recall such a highly dramatic scene?

HW: No. Besides, I hardly ever went to the village to shop, or to the baker.

FH: But it could be that, for once, you did go to the baker, to fetch a cake, possibly, because Frau Luchaire had quite unexpectedly come to visit. That would be conceivable. However, I can hardly imagine that in the presence of Frau Einstein and a strange lady you would have used such a vulgar expression as sonovabitch for the baker. Even if you were outraged at his comment—assuming that all this actually happened—your choice of words would surely have remained under your control.

HW: The expression sonovabitch was never part of my vocabulary, not then or later.

FH: I can believe that. Quite apart from that, such an incident would not be completely impossible in the prevailing political climate, even if you cannot recall it today. It is a psychological finding that unpleasant experiences can often be completely forgotten or repressed. Let us leave it at that. I would like to return to Natalia Saz and her memoir, *Novellas of my Life*. She writes that on a visit to Caputh in the spring 1931, Einstein had worn a jacket of pure wool that, he proudly said, had been knitted by his "Ilse"—meaning Frau Elsa—with her own hands.

HW: In cool weather, Herr Professor liked to wear a pullover or a woolen jacket. But that Frau Elsa supposedly had knitted it, that's news to me. I never saw her knitting.

FH: Did Einstein then receive his visitors in a wool jacket or pullover?

HW: If it was an official visitor, Herr Professor always dressed appropriately. That can also be seen from the photo with Tagore: On the terrace, where he usually sat in his warm-up suit or bathrobe, Herr Professor is seen to be dressed in a business suit when he and his Indian guest were photographed, and he had even put on a tie.

FH: Incidentally, a DDR author who is famous today also visited Einstein in Caputh: Anna Seghers. She recently described in a weekly magazine how she traveled to Caputh one summer in order to ask him to give a lecture at the Marxist Workers School in Berlin, known as the MASCH. She remembers him well, but what she recalls even better was the cucumber salad that was served at the midday meal and that he relished it greatly. She remarks on Elsa's friendly smile and that she tried to keep her husband from committing himself to the lecture.

HW: Anna Seghers was not a name I knew at that time.

FH: But to this day, she fondly remembers the cucumber salad that you had presumably prepared. By the way, Einstein did give the requested lecture at the MASCH. It intrigued him, evidently, to explain the results of his research to ordinary people in this "Workingmen's School." On October 26, 1931, he spoke in the auditorium of an old school building in the North of Berlin, addressing more than three hundred workers and unemployed persons, on the subject "What the Worker Must Know about the Theory of Relativity." According to participants in the event, the lecture was followed by a lively discussion about philosophical and political issues.

HW: So many visitors came to Caputh that I did not always learn their names, or else have forgotten them long ago.

FH: Arnold Zweig is among the well-known authors who looked up Einstein in Caputh. He told me so himself in 1961, when I asked him about his encounters with Einstein. He had not been to the city apartment, but he did travel to Caputh once in order to obtain Einstein's signature for some joint appeal or other. Since we are talking about well-known writers of those years: Einstein was also personally acquainted with Berthold Brecht, though they met only occasionally, at receptions or dinner parties. This is what Helene Weigel wrote to me, when I asked her about possible meetings between Brecht and Einstein.[46] She thought it unlikely that Brecht came to the Haberlandstrasse or to Caputh.

But now for something entirely different: You mentioned to me that once there were some uninvited guests in the summer house.

HW: It must have been in the earliest time. When we came to Caputh in the spring, we could see that vagrants had been there. They must have entered through the furnace cellar but they stayed only in Herr Professor's room. Blankets were all over the room, and we discovered to our dismay that they had left bed bugs behind and an exterminator had to disinfect the room.

FH: Were any valuables or other articles missing?

HW: No. There were actually no valuables in the summer house. The silver cutlery used for large parties remained in the city apartment, for in Caputh everything was supposed to be more rural. Portable items that we did not take back to the city apartment when we returned to Berlin in the fall were left in Caputh, in pottery master Wolff's attic. In the spring, we then brought them back up to the house.

FH: The house was insured, incidentally. According to the fire insurance policy that I was able to see, the house was built with "Orion pine lumber," the high-grade timber of American pine that is particularly hard and long-lasting. It was very widely used in timber construction in Germany.

HW: That reminds me of something related to Caputh. My mother, who often visited me there, was very interested in the flowers on the property. She liked certain high bushes with yellow-brownish flowers, called "Sonnenbraut" [*helenium*], which were quite rare then but can now be seen in many gardens. These were immediately ordered for my mother, from the palace gardener of Sanssouci, and were shipped to Lautawerk.[47] My mother planted them in her garden and told everyone proudly: "They come from the palace garden in Sanssouci and Herta sent them to me, thanks to Herr Professor Einstein's mediation."

FH: Did the Einsteins employ their own gardener?

HW: In the beginning, it was Meier, the unemployed typesetter whom I mentioned before, who acted as gardener and later it was another resident of Caputh. He is also the one who ordered flowers from the palace gardener and planted them in the flower boxes on the lower terrace and in the garden room.

FH: One last question regarding Caputh: Was the fireplace in the living room intended as an architectural decoration, or did it fulfill its real purpose? Fireplace fires are generally considered to be decorative heating, because they are so inefficient.

HW: When it became cool outside, we put a few birch logs in the fireplace to make the living room cozier. The fire also provided a lovely shimmer. It gave a lovely flickering light when one sat there in darkness. I can still see that picture clearly before my eyes.

FH: Did the Einsteins often sit in front of the fireplace?

HW: Yes, particularly in the late autumn when the evenings were already long.

FIFTH CONVERSATION: THE HOUSE SEARCHED AND PLUNDERED—INTERROGATION

FH: In the course of our conversations you spoke repeatedly of Einstein withdrawing to his study to cogitate about something or to write. It seems fitting to ask what he was pondering during those years, and what he may have written.

In theoretical physics, his main work area, Einstein wrestled at that time with the requirements of a unified field theory. Its system of equations was to describe the gravitational field, as well as the electromagnetic field, and thereby provide a deeper understanding of nature. He submitted the first version to the Proceedings of the Academy in 1925 and wrote a second one at Plesch's country estate in Gatow, in January 1929. It soon became clear, however, that the time was not yet ripe for such an all-encompassing physical theory.

The other theme that occupied Einstein's mind from the middle of the 1920s was quantum mechanics. In 1905 Einstein had made a crucial contribution to quantum physics with his theory of light quanta, and he made other contributions to its foundations until the early 1920s, when he found himself in an increasingly bitter disagreement with the others working on quantum mechanics. During the Solvay conferences of 1927 and 1930, a stubborn dispute surfaced between Einstein and Niels Bohr, who was the leading proponent of the so-called Copenhagen School, along with Werner Heisenberg, Wolfgang Pauli, and other young physicists.[48] Einstein was unable to counter Bohr's arguments but would not concede. He was unable to accept the probabilistic interpretation of quantum phenomena as the final answer. For him, it was merely a stopgap measure, and he stuck to this view to the end of his life. More about this is found in my books *Einstein and His World View* and *Pioneers of the Atomic Age*, as well as in the short biography *Albert Einstein*.[49]

Einstein had manifold scientific interests and during the years 1927 to 1932 he occupied himself with the most diverse problems, as evidenced by the over a hundred scientific and general publications he produced during those six years. From among them, I would like to single out his commemorative essays about Johannes Kepler and Isaac Newton. His obituaries for the American experimental physicist Albert A. Michelson and the German foreign minister Gustav Stresemann also deserve mention.[50] Besides the printed writings of Einstein, there exist two remarkable spoken texts from those years: His speech at the opening of the broadcasting exhibition in Berlin (August 1930) and the recording "My Credo" [*Mein Glaubensbekenntnis*] that Einstein made for the German League for Human Rights in autumn 1932, which is a rarity today. I have published transcriptions of both of these acoustic documents in the journal "Naturwissenschaften," in 1964 and 1966.[51]

Einstein's remarks "On the Death of Lenin" belong to the same period. Although written earlier, they were not published till 1929, in the birthday volume *Occasional Pieces by Einstein* mentioned before. They reveal clearly the great scientist's humanitarian cast of mind. He wrote, "I honor Lenin as a person who employed all his energy, and sacrificed himself to achieve social justice. I do not consider his method appropriate, but one thing is certain: Men like him safeguard and restore the conscience of mankind."[52]

In those years, particularly after 1930, Einstein composed and signed numerous declarations and appeals, most of them urging the formation of a united front against the emergent German fascism. In autumn of 1932 he issued such an appeal together with Heinrich Mann, Arnold Zweig, Käthe Kollwitz, Ernst Toller, and other German antifascists.[53] In the summer of the same year he supported the International Congress against Imperialist Wars in Amsterdam, as a member of the Invitations Committee. On that occasion he met with the British philosopher Bertrand Russell, as well as with progressive, antimilitaristic writers like Romain Rolland, Upton Sinclair, Heinrich Mann, Martin Andersen Nexø, Henri Barbusse (author of the antiwar novel *The Fire* [*Das Feuer*]), and Karl Kraus (publisher of the periodical *Die Fackel*)—to mention a few.[54] In November 1932, while residing in Caputh, Einstein, along with his French colleague and friend Paul Langevin, tried to organize the most determined antiwar intellectuals for collective resistance against fascism and militarism. But fate took its course.

Einstein was staying in Pasadena as a visiting lecturer when, on January 30, 1933, German financial interests thrust the Hitler clique into power. Documents in the Central Archive of the Academy of Sciences of the DDR show that at first Einstein did not attach enormous significance to the change in government. Early in February 1933, he was negotiating, while still in Pasadena, with the Academy of Sciences in Berlin regarding a revision of his employment status. He requested the Academy to grant him, each winter, a recurrent unpaid leave of absence of several months, so he could take up his duties at the newly founded Institute for Advanced Study in Princeton. For the remainder of the year he wished to work mainly at the Academy, as before. The administration of the Academy agreed to this arrangement, which implied Einstein's continued presence in Berlin. When it became evident, however, that the fascist "scourge," as Einstein called it, was spreading in Germany, he decided not to return to his native land. He did, however, travel to Europe once more, when he and his wife, Elsa, stayed in Belgium, as guests of the royal couple, in Le Coq Sur Mer, a small seaside resort near Ostend. From there, Einstein traveled twice to England to lecture, and in the fall of 1933 he moved to the United States to take up his agreed-upon activities in Princeton. In March 1933 he announced his resignation from the Prussian Academy of Sciences in a politically demonstrative way and renounced his German citizenship.

FH: Were there any consequences in the Haberlandstrasse when the Nazis came to power?

HW: At first there were none. At least, I can recall nothing of that kind.

FH: Did you go out to Caputh again in the spring of 1933?

HW: I had no reason to do so.

During those weeks and months, not a few of the Einsteins' friends felt compelled to emigrate from Hitler's Germany. Among them was Max Born who, as a Jew, had to give up his professorship in Göttingen. Erwin Schrödinger, who used to go sailing with Einstein in Caputh, was not obliged to emigrate, but he gave up his professorship at Berlin University voluntarily and accepted an appointment in Oxford because he wanted nothing to do with the fascists.[55] The atomic physicist Lise Meitner, whom Einstein liked to call "our Madame Curie," was barred from teaching at Berlin University—just one of about two hundred and thirty faculty members of that university who were dismissed on racial or political grounds in that summer of 1933 and thereafter.[56]

Among Einstein's medical friends, János Plesch may have been among the earliest emigrants. Toni Mendel, who was also of Jewish heritage, had to give up her beautiful property in Wannsee and lived in her country house on the Zürcher Lake in Switzerland after 1932. This is also where her other close relatives found their first refuge, after fleeing from Hitler's Germany. Shortly before the Second World War broke out, she moved to the United States, where she also died. It is said that Einstein, who survived his friend by several years, was in touch with her in the United States, in person and by letter.[57]

Toni Mendel, during her last years in Berlin, has been described as a slightly built, cultured lady, who did not dress glamorously, about the same age as Einstein, and at the time already widowed. She was Einstein's companion at concerts and at the opera, and he quite often spent the night in her luxurious villa on the Wannsee. In its music room, which afforded a splendid vista of the lake, Einstein occasionally played the piano so loudly at six o'clock in the morning that it woke up the other occupants of the house: A sort of pianistic companion piece to his nocturnal violin playing in the Haberlandstrasse.

It is reported that Einstein conversed about philosophical questions with this cultivated, sophisticated woman, who was apparently an ideal partner for him. Einstein's express wish that after his death, all his letters to Toni Mendel be burned by her heirs—which did happen—allows one to conclude that very private matters occurred during their frequent encounters. In view of the deep insight into the psyche of important personalities that intimate letters can reveal—Ernst Haeckel's letters to Frida von Uslar-Gleichen ("Franziska von Altenhausen") serve as examples—one must regret the loss of these documents.[58]

One more comment that may also be of interest to historians of medicine: Toni's son-in-law Bruno Mendel, a nephew of her [late] husband, was an independent scientist and had built and equipped a private research laboratory on her Wannsee property. He was a physiologist and an MD and in his research on the causes of cancer and on cancer therapies he was personally counseled by the Nobel Prize recipient Otto Warburg.[59] Einstein took a lively interest in his experiments using white rats, and on his visits to the Wannsee establishment he often sat on a lab stool, marveled at the precision [physiological] instruments, and offered technical suggestions.—for example, for the construction of an electrically heated bath tub that was needed to facilitate "heat-therapy" for the eradication of cancer cells. This is further evidence of the diversity of Einstein's technical interests.

HW: Now it is my turn to ask you something: How did you learn this detailed information about Toni Mendel and what happened in Wannsee?

FH: Through a lucky coincidence. Several years ago, during one of my Einstein lectures, I made the acquaintance of an elderly colleague who had been employed from 1925 to 1933 as a laboratory assistant in the private laboratory in Wannsee, and recently I questioned her more exhaustively about her memories of Einstein. She also told me that, during his stay in Berlin, Tagore spent several days in the villa on the Wannsee.

HW: I heard nothing about that at the time.

FH: Following this detour, let us return to the chronology of events in the city apartment and the summer house. I want to begin by clarifying something I found in many Einstein biographies, whose authors were even able to cite remarks Einstein made on his return journey to Europe, in March 1933. These remarks were reprinted in a volume of his writings *Einstein on Peace* [*Einstein über den Frieden*], together with a commentary that gives the impression that the described events actually took place.[60] Einstein explained that, in the past, his summer house in Caputh was often honored by the presence of guests, who were always welcome. Nobody had cause to enter by force. That this had now happened, was another example of the arbitrary acts of violence taking place in fascist Germany. Einstein's comment was occasioned by a report from foreign correspondents in Berlin that his house had been ransacked by the Nazis, who suspected that weapons were hidden there. Based on this newspaper report, Philipp Frank wrote in his Einstein biography, "Einstein's villa in Caputh was searched by the political police. It was suspected that the Communist Party had hidden a weapons cache there." Antonina Vallentin joined in with this supposition but garnished it in her particular style: In their search for hidden weapons—she writes—the garden soil was "dug up methodically," thanks to an obliging neighbor who hurried to put shovels at the disposal of the "Hitler henchmen." An American Einstein biography states that a troop of SA men surrounded the Einstein house and then searched it for weapons but found only a bread knife! Did you hear of such a raid in the spring of 1933? Although you were staying in the city apartment, you would surely have been told about it by Dr. Kayser, who was in touch with Caputh in those weeks—as is known from archival evidence.

HW: I know nothing of such a break-in.

FH: After interviewing the owners of neighboring properties, it turned out that the newspaper report was a hoax. One could simply ignore this Einstein legend, as one among dozens, if it did not defame his former neighbors, who are alleged to have been zealous helpers of the SA. Even though the Nazis were capable of every infamous act, they did not begin to take an interest in Einstein's summer house until May 1934—two months after the physicist and Nobel Prize recipient was stripped of his citizenship, together with Johannes R. Becher and other German antifascists whose names were listed in the government gazette [*Reichs- und Staatsanzeiger*]. This is when the pertinent authorities took steps to confiscate the Caputh house, which was at the time—from April 1933 to April 1935—rented and utilized by a Jewish agricultural school. After a prolonged back and forth, the Gestapo ordered that the house be appropriated for the Prussian State in 1935, in accordance with the Nazi laws for the confiscation of property owned by Communists and enemies of *Volk*, or the state. I have documented the specific stages of the Nazi foray against Einstein's house and lot in the article I mentioned before.

HW: I knew nothing about these events at the time.

FH: I reconstructed them from documents I was able to examine in various archives. The biographical literature's accounts of what happened to Einstein's sailboat are also often incorrect. Thus, Antonina Vallentin has it that Einstein, after emigrating, rediscovered "his beloved sailboat" on a lake near Montreal. That is as much a legend as is the widespread assertion that, in the spring of 1933, the Nazis offered a reward of 20,000 or 50,000 Marks for Einstein's capture.

But there is another matter in need of clarification that I want to bring up, because you are mentioned in it, though not by name. An American Einstein biography states that immediately after landing in Belgium—that is, at the end of March 1933—Frau Elsa telephoned the Haberlandstrasse apartment and that the "housemaid" told her through her tears that Marianoff and Margot had fled to Paris and that Ilse and her husband had already crossed the Dutch border. Do you recall this telephone conversation, which I consider most unlikely under the prevailing circumstances.

HW: No. Besides, at the end of March there was nothing alarming to report from the Haberlandstrasse, certainly not "through my tears." Dr. Marianoff did not leave for Paris until the beginning of April, and Margot followed him shortly afterward. The Kaysers did not cross the Dutch border; they remained in Berlin for several months, and I worked for them for four or six weeks in the summer of 1933.

FH: So this is another touching tale about Einstein and more evidence that such reports should not be taken too seriously. Let us now leave the domain of journalistic-biographical fantasy and return to the firm ground of political reality! You told me once that, shortly after the Marianoffs left, the criminal investigation police searched the apartment in the Haberlandstrasse. How did this take place?

HW: It must have been the beginning or middle of April 1933, quite early in the morning, between six and seven o'clock. I was alone in the apartment and was still asleep when the doorbell rang. Three or four men in civilian clothes stood in front of the door. I had opened the door but left the security chain in place because I wasn't properly dressed yet. I had been asleep, after all. The men said "*Kriminalpolizei*," and I answered, "Your identification, please!" They showed me their badges and then I opened the door and let them in. They did not ask after Herr Professor, however; they only wanted to know where Dr. Marianoff was. I said that he and his wife had gone on a trip. They then wanted to know which room was his, and since the Einsteins were in America he had stayed in Herr Professor's room. They searched through everything and asked me when the Marianoffs had left. One of the officials stayed with me in the kitchen, and the others left to speak with Frau Dr. Kayser, presumably to see if Dr. Marianoff and Margot were hiding there and to check if my statement that the two were traveling agreed with Frau Dr. Kayser's. One of the policemen stayed with me in the kitchen to keep me from making a phone call, but he was only interested in Dr. Marianoff, not at all in Herr Professor. He was a young man, sitting with me in the kitchen, and since I was having breakfast I offered him a cup of coffee. But he just kept asking me about Dr. Marianoff.

FH: Did they search all the rooms? Including the tower room?

HW: They only gave a glance to the other rooms but did not rummage around in them. They didn't go upstairs, to Herr Professor's study, and I assume that they did not know anything about it. Mostly they looked around the room that Dr. Marianoff had occupied.

FH: How long might the search have lasted?

HW: Maybe half an hour or a little more. Then one of the men sat down in the kitchen with me, as I said. A little later, a telephone call came, and he left, as well.

FH: This happened in the middle or end of April 1933. You told me that several weeks later the apartment was ransacked by uniformed men, under the pretext of a house search.

HW: Frau Dr. Kayser was in the apartment at the time, as well as Fräulein Dukas. I am pretty certain that Fräulein Dukas was present, because the household was being dissolved, and she was responsible for packing the books and papers, I was certainly not alone in the apartment with Frau Dr. Kayser; there were three of us, I am quite certain.

FH: When was that?

HW: It must have been the end of May or the beginning of June. More likely the end of May. We were busy packing everything up.

FH: What was the time of day?

HW: Toward evening, because Frau Dr. Kayser always came in the afternoon.

FH: What happened?

HW: Several men in uniform rang the doorbell. Since I was not alone, I was not as careful as I had been the first time. I did not ask for identification, nor for a search order.

FH: Did they expressly say that they had to carry out a house search?

HW: Yes. They said it was a house search. Frau Dr. Kayser had to open all the cupboards for them and then all three of us were told to sit in the library and stay there. We were expressly told that. While we were sitting in the library they had me give them the keys to the building and to the elevator, so they could leave the building, which was always locked. At first we heard them running around and then it became very still. I said, "I am getting a little hungry; I am going to the kitchen to get something to eat." Frau Dr. Kayser was very frightened, however; she warned me and asked me not to go out. But I was hungry and wanted to see what was going on. When I entered the hall, I saw that nobody was in the apartment, and we noticed right away that the carpets were gone.

FH: The carpets? The ones Chaplin had called old and threadbare?

HW: Those carpets were now gone. We had already rolled them up to get them ready for putting them in storage. The pictures were taken, too, but which pictures I naturally can no longer recall. But I do recall very well that my overcoat that had hung in the guest room had also disappeared. I often slept there when nobody else was in the apartment so I could hear everything better. That's why the coat was hanging there. They probably used it to wrap up the things they took: the tea pot, the coffee pot, and other silver objects. They had taken from the cupboards whatever appealed to them. They were evidently out to loot the place, even though they said that they came to search the house. Since there had actually been a house search a few weeks earlier, we had believed them.

FH: That must have been quite upsetting for all of you.

HW: Of course. When we discovered the robbery, I immediately ran down to the porter, and he told me that the men had loaded everything on a truck and had driven away. I then went to the police and reported the incident the same evening. I was really very upset. Since they had taken my keys with them, I was afraid that they might return. I asked at the police station what I should now do, if someone could come with me. We then agreed that a woman friend of mine would come to the apartment with me, and she stayed the night with me. Fortunately, I had a spare key.

FH: What did the police do?

HW: Supposedly, they knew nothing about it at the police station, but they also did hardly anything. I kept on saying, one could surely find out who the people were, they were in uniform, after all. But it was of no use.

FH: What uniforms did the robbers wear?

HW: As I remember, they were SA uniforms, but perhaps not all of them.

FH: How many were there altogether?

HW: At least five or six.

FH: How long did the operation last? I mean, from the time you and the two other women were sent to the library to when you emerged to get something to eat in the kitchen.

HW: We sat at least one and a half to two hours in the library.

FH: Did the thugs also go upstairs to the tower room?

HW: They were only in the apartment; I assume that they knew nothing of that room.

FH: Did they wreak havoc in the apartment? Did they destroy furniture or other things?

HW: Nothing was destroyed or smashed. They just stole things they liked and vanished with them.

FH: One sometimes hears the view that the Nazis smashed everything into little pieces in Einstein's city apartment, which does not correspond to the facts. Chaplin's supposition that German fascists had chopped the piano in the Biedermeier room into firewood is also untrue. The instrument arrived intact in Princeton, and Einstein often played on that Bechstein grand in his last years, or was accompanied on it by visiting musicians. Chaplin's notion fits into the long-standing cliché of German fascism and is appropriate for the stage, as Paul Dessau's opera *Einstein* demonstrates.[61] But, particularly in the first years, the Nazis employed methods other than those used in *Kristallnacht*, in 1938, as was shown by the Soviet film pioneer Michail Romm in his impressive documentary film

Ordinary Fascism.[62] The roguery by the SA in the Haberlandstrasse that you witnessed is further evidence of this. By the way, one must say that you really acted quite fearlessly, and when Einstein, in retrospect, called you "courageous Herta," he may well have had your undaunted conduct during that incident in mind. He must have heard of it.

HW: I would assume so, because Frau Dr. Kayser and Fräulein Dukas were present.

FH: What happened to the furniture and other household furnishings that had not been stolen by the Nazis?

HW: We shipped them to a warehouse of the shipping company Silberstein, in Schöneberg.

FH: Did the furniture transport take place at night, so it would draw no attention?

HW: Oh dear, no; it was a quite ordinary move. Everything was loaded into several moving vans and taken to the warehouse. My cousin helped me with the packing. Frau Dr. Kayser decided the destination of each item.

FH: Many years ago, Helen Dukas informed me that Einstein's papers and manuscripts went via the French Embassy to Paris, as diplomatic baggage not subject to inspection, and were forwarded to Princeton from there. All of Einstein's books in the city apartment also went to Princeton, as did the household goods, and they were shipped under Dr. Kayser's name. There is documentary evidence that Kayser also rescued many items in Caputh, or tried to. I have read submissions and letters by him and his attorney that raise objections to the confiscation of Einstein's sailboat and, later, to the misappropriation of his house and the parcel of land. Naturally, all in vain. As you know, Kayser and his wife emigrated to Paris in the summer of 1933. After she died, he stayed for some time in the Netherlands, before he, too, finally emigrated to the United States. There he taught at a university, as professor of German literature. In 1946, Rudolf Kayser published a book about the philosopher Spinoza, with a foreword by Albert Einstein, who revered this sage.[63]

HW: I did not know that. But I knew the couple very well, not only from their frequent visits to the city apartment and Caputh. Once, I also did the cooking for them for four weeks, when the Einsteins were away from Berlin for some time and the Kaysers' housekeeper was ill in hospital.

FH: According to your statement in the transcript from September 5, 1934, which we will talk of directly, and Frau Einstein's letter of recommendation, the household in the Haberlandstrasse was dissolved on June 1, 1933.[64] Was the apartment emptied completely then, or were objects of no particular value left behind?

HW: The apartment was completely empty. It was agreed that I would be allowed to take the so-called daughter-room furniture, which was painted bright green and was actually intended for a young girl's room, as was then the custom. There were altogether two cupboards, a bed, a round table, and also the smaller table on which Margot used to do her modeling; also, a vanity table and a small ladies' desk. The things that were not particularly valuable were not sent to the warehouse to be shipped to America. I was supposed to take them, and everything else was shipped. Nothing was sold. In Caputh, everything stayed where it was.

FH: Except for the items stored at pottery master Wolff's house for the winter, which friends of Dr. Kayser brought to Berlin. This is known from the town archive's records in Caputh, and it was confirmed by the local inhabitants. The record shows that the upper neighbor, Frau Bornemann, had sewn sixty-six buttons on two American comforters that were also picked up. Sometimes even buttons end up in an official report.

HW: I knew nothing of all that and also had nothing to do with it. But I would like to mention something about the time I spent at Dr. Kayser's. The Kaysers owned a large tomcat that always sat on the desk when Herr Dr. Kayser was writing. He was brownish in color, with stripes like a tiger. Oh, that was a splendid beast! And was he spoilt! I always had to buy kippered tench [*Tinca tinca*] for him; he did not like any other kind of fish. At mealtimes, he also sat at the table, a napkin was spread out on the table, and the tomcat sat on a stool and placed his front paws on the napkin. He also ate in a very decorous manner and didn't make a mess.

FH: That reminds me of a custom that the chemist Wilhelm Ostwald, whom I mentioned before in our conversation, introduced in his home in Grossbothen, near Leipzig.[65] Their cat always sat at the table on equal terms with the family, often to the dismay of guests, who were not used to such table customs. When I worked in the Ostwald archive in 1956/57, I was shown the little basket chair that had been specially made for this purpose. I don't suppose that your Peter in Caputh sat at the table, however.

HW: Oh, dear me, no! Peter always ate with me in the kitchen, and he was not spoilt. Stray cats are not like that.

FH: Since we are talking of cats: Einstein was apparently deeply involved with cats in Princeton. Since he considered himself a sailing devotee in his Berlin days, he must have been something of cat devotee in Princeton. The biography that was co-authored by Helen Dukas and is, at least for Einstein's last decades, apparently devoid of legends, describes how he used to console his cat, which became very depressed whenever it rained, with the words, "I know what is bothering you, dear girl, but I really don't know how to turn it off." The same book also describes how he once took a very long way home in order to see the newborn kittens of a collaborator's cat. When he noticed that their way took them to a neighborhood where many institute colleagues lived, he said to his collaborator, "Let's walk a little faster. I have declined the invitations of so many people here, I hope they won't notice that I am here to visit your cats."

HW: I read that, too, and was much amused. Margot wrote to me after the war that she once had thirty cats at the same time in Princeton, whereupon "Father Albert" had said that he now had enough.

FH: Who could disagree with that? Incidentally, Lenin also had a cat in his two-room apartment in the Kremlin. It was the darling of the family and used to jump on his shoulder and liked to snuggle on his lap. Clara Zetkin describes it lovingly in her recollections of Lenin, as did other visitors, as well. There are documentary films that show him petting his tomcat.

But following this amusing digression about cats and kittens, let us return to you. What course did your life take after your employment in the Haberlandstrasse came to an end?

HW: Immediately afterward, I took my paid annual vacation and then I was employed at Dr. Kayser's home for four to six weeks, until that household was dissolved, as well. That had been agreed upon. In the late summer of 1933 I began a new job in Moabit, in the household of a physician, a surgeon.

HW: That is where the police tracked you down in the summer of 1934; you were to be questioned by police investigators, at the instigation of the Gestapo. The immediate motive for the hearing was that Einstein's house in Caputh was to be the expropriated. Einstein had allegedly supported Marxist aspirations and had held secret negotiations behind locked doors with Communist functionaries, as well as with other enemies of the state. Since there was no written evidence to serve as the basis for the expropriation, it was planned—so reads the record—"to discover, by skillfully interrogating former domestics, who the persons were that visited the Einstein house." So it seems that the Gestapo had assigned a key role to you in the pilfering of the Caputh house. They wanted to employ you as a star witness for Einstein's subversive activities against the state. May I ask you to describe in chronological order what happened at the police hearing on September 5, 1934.

HW: The police first contacted my parents in Lautawerk to find out my address, without providing any reasons. Soon afterward, I received a summons in Berlin, to appear at police headquarters on the Alexander Platz on September 5, 1934. No reason for the summons was given. It caused me many sleepless nights because I did not know what they wanted with me. I breathed a sigh of relief, however, when I was called into the room and saw that the folder on the table was labeled "Einstein." So, it would not be about me, but about Herr Professor, and he was safely far away. I felt tremendously relieved. I had to give my personal data, and then the investigator, who had already been sitting at the desk when I entered, asked me if the Einsteins still owed me money, whether they had been good to me, or if they had possibly mistreated me. The official asked all his questions in the expectation of getting negative answers that could incriminate Herr Professor.

FH: How long did the hearing last?

HW: About half an hour or a perhaps a little more. Then I had to sign the transcript and could leave. I never heard anymore from the Kripo [criminal investigation police].

FH: A few years ago I was able to examine the transcript from the hearing in one of our archives, and since it has a bearing on matters mentioned before I would like to reproduce it verbatim. This is your testimony before the National Police Department II, 1b, Berlin, on September 5, 1934:

> I worked for Professor Dr. Albert Einstein, Berlin W30, Haberlandstr. 5, from 15.6.1927 to 1 June 1933 as a domestic employee. In the course of my stay there, very many personages came to visit, among them, Gerhad [sic] Hauptmann, Professor Planck, Professor Ehrmann, Professor Lichtwitz, Professor Maier, Musikdirektor Kleiber and many others. There was a lively coming-and-going in the house at all times. Many foreign personages visited, as well. Many pacifists and Zionists also visited there. During such visits, I frequently had the opportunity to listen to conversations and I heard nothing suspicious in them. In those days, it would not have seemed remarkable to me for political conversations to take place. The conversations with the visitors took place openly, i.e. not behind locked doors.
>
> Einstein with his wife had left for America already in December 1932 and as a result of the [government] overthrow in Germany, he has not returned here. As far as I know, Einstein is supposed to be in California and to be active professionally there.
>
> The summer home in Caputh was closed up every year in October, sometimes November, when the move to the apartment in the Haberlandstr. was made. This is also what happened in December [19]32, when E. went to America and did not return from there.
>
> I am unable to provide any information of a political nature about E. and his daughters. The parcel of land in Caputh belongs, as far as I know, to the children of E. The daughters are the children of Frau Einstein from her first marriage.

So much for the text of the transcript, then followed by the usual abbreviations for "was read to—was agreed to—signed by," and, on the left, your signature. On the right side is the signature of the investigator who questioned you. One comment: Professor Maier surely refers to Einstein's last assistant in Berlin, Dr. Mayer. But who is Professor Lichtwitz?[66] I have never heard you mention that name when you spoke of Einstein's visitors, nor could I find him in any reference work.

HW: Who that was, I cannot recall. It could be that someone mistakenly heard it when the transcript was made.

FH: Your interrogation was supposed to provide formal, legal grounds for confiscating the Caputh property. As it turned out, however, it drew a complete blank. This was also clear to the officials charged with the confiscation, for it says in one of the documents: "Evidence that the property was used for subversive activities against the state could not be obtained by interrogating the domestic employee." What the Gestapo wanted to hear from you, they failed to get. Although you were well aware of the Einsteins' political point of view—I have the scene at the polling place in mind—you avoided all mention of it and professed to be ignorant. Any mention of Frau Einstein's remark would have been grist in the Gestapo's mill. Although they wanted to hear names of KPD [Communist Party] functionaries with whom Einstein supposedly plotted against the state, the SPD [Social Democratic Party], who received the Einsteins' votes, was also a Marxist party to the Nazis; by September 1934, it had also been outlawed long before. The transcript makes no mention at all of visitors to the summer house, which was, after all, the object of the investigation. The questioning deals more with the city apartment. It struck me, by the way, that out of the large number of Einstein's guests, you selected only a few. For example, why did you specifically name Gerhart Hauptmann and Erich Kleiber?

HW: At the time, Gerhart Hauptmann had, supposedly, made some objectionable comments regarding the Jewish question and sympathized with the Nazis. I did not like that at all. I named music director Kleiber because he belonged to those Jews who now repudiated Einstein because they held him chiefly responsible for the persecution of Jews.[67]

FH: In April 1933 Frau Elsa Einstein commented on this issue in a letter to Antonina Vallentin from Belgium: "What makes my husband's fate so tragic is that all the German Jews hold him responsible for the horrors they are now experiencing. They believe that his conduct has led to their repression, and in their narrow-mindedness made it their watchword to turn away from him and to hate him.[68] As a result, we have received more hate-filled letters from Jews than from Nazis. And, if truth be known, he sacrificed himself for the Jews! He was not intimidated, and he did not fail them. Is it not a tragedy that the same people who deified him now throw dirt at him?" In another of her letters, Frau Elsa wrote to Antonina, "The Jews [in Germany] are so terrified and misjudge their situation so badly, that my husband cannot reach them. They have removed all his pictures, or burnt them."

HW: As I recall, music director Kleiber was also among these Jews, and that is why I named him. He visited Herr Professor often, and they used to drink tea together. I found it not nice of him to behave that way later. But, as I said before, I was questioned with particular emphasis on whether the Einsteins still owed me money and how I had been treated in the family.

FH: The Gestapo would have loved to hear that the "Aryan" housekeeper had been mistreated by the "Jew Einstein"—as he is referred to in several reports. That is why the police official kept asking such questions.

HW: But then, when he noticed that I had only good things to say, he gave up asking questions along those lines. I said to him, "If I had been treated badly, I would hardly have remained in that household for six years!" To this he responded, "Well, yes, formerly one would say: Hosanna, today one says: Stone them to death!" This remark has stuck in my memory, I remember it word for word. I was impressed that he dared to say that.

FH: The investigator could hardly have been a fanatical Nazi. But it was all in vain! The confiscation of the Einstein house by the Gestapo was a done deal, and it was concluded despite the lack of evidence of Einstein's subversive activities on this property. What was the reaction of the physician whose household you kept, to your summons? Under the prevailing conditions, it could hardly have been a matter of indifference that his domestic employee was summoned to police headquarters for a hearing.

HW: I told him everything afterward, and he was very understanding. His wife also calmed down when she heard why I had been summoned to police headquarters. Both of them had, at first, been very concerned about me.

FH: How long were you employed in Moabit?

HW: Until they issued a law that forbade "Aryan" employees from working in Jewish households. Then I took a job in a home for Jewish children near the Rosenthaler Platz. I was allowed to work there because it was a public facility, not a private household. I stayed in that home until I married in October 1936, when I gave up my career as a domestic employee.

FH: Did you receive mail from the Einsteins following their emigration?

HW: Yes, I got frequent letters from Margot, at least in the beginning. But there was nothing political in them, for she had to count on the letters being censored. After my hearing by the criminal investigation police in September 1934, it often happened that letters to me were monitored. Not only letters from abroad, but also letters from my mother in Lautawerk to me. They had been opened and sealed again; that was plain to see.

FH: They probably suspected that they might find clues that suggested a continuing connection to the "Jew Einstein" and his family. The Gestapo had your precise address and kept a watchful eye on you. Anybody who had worked for a Jew for six years and was now again employed in a Jewish household had to appear suspicious to these people.

HW: That reminds me of a particular occurrence. Margot sent me a piece of jewelry from Paris, but it arrived totally crushed. One could see that it had not been damaged accidentally in transit but that it had been purposefully, wantonly crushed in order to make it worthless. It was at the time—1934—when I learnt of Ilse's passing, and the jewelry was meant as a personal token in memory of Frau Dr. Kayser. Margot had nursed her sister until she died.

FH: In order to make the connection to the time after the Second World War, when you corresponded with the Einsteins again, I would like to ask you for a brief account of how you survived the war years.

HW: I was in Berlin until 1943. My husband was called up for military service. When the air raids became increasingly heavy, I went to Lautawerk with my boy and stayed with my younger brother. In the end, we found refuge with some kind people in the mountains of Northern Bohemia [*Erzgebirge*]. When the war ended, I returned to Berlin, where our old apartment had miraculously been left standing. I worked as a *Trümmerfrau* [cleaning up rubble], like many other women in Berlin. Then, in 1951, the National Chamber of Commerce was looking for saleswomen and I signed on, initially for four weeks, but they would not let me go again. And so I worked there for sixteen years, until my retirement: as sales person, sales supervisor, and cashier, and I was honored as a political activist [*Aktivist*]. My husband returned from an English prisoner-of-war camp in 1948 and got a job in a publicly owned business. Our son studied civil engineering and is now working as a project planner.

FH: Please tell me a little more about your correspondence with the Einsteins after the Second World War.

HW: I hesitated for a long time before writing to Herr Professor because I did not want to appear as a supplicant, but in 1947 I did write him for his sixty-eighth birthday on March 14. I wrote how I had survived the war years and how I was. Just when I wondered if my letter had by now reached Princeton, an airmail letter from Margot arrived. Margot wrote of how often they had all talked about me and how they had worried about me. And enclosed in her letter was a handwritten letter from Herr Professor.

> Dear Herta!
>
> You should have seen the jubilation when your letter arrived. How often had we asked ourselves what might have happened to our dear courageous Herta? And we were concerned that we heard nothing for such a long time. After all, you lived through so many diverse experiences together with us, in the now vanished Haberlandstrasse and in Caputh. The Russians now have the little house and a Nazi got the plump sailboat. I hope that things will soon get better for you with regard to the wretched food supply.
>
> Many warm good wishes,
>
> Your
>
> A. E.

FH: There are two comments to make regarding Einstein's letter. The remark "The Russians now have the little house" must not be understood to mean that the Soviet occupation army had ever taken possession of the house and land. The record shows that in the summer of 1945 the Soviet commander ordered the mayor of Caputh to return the house to its original state immediately, in case Einstein wished to return. Unfortunately, Einstein did not learn of this. Probably, all he meant to say with his comment is that his erstwhile summer house was now—i.e., in April 1947—in the "Russian" occupation zone.

As far as "the plump sailboat" was concerned, he was evidently misinformed. I found this account in the records of Caputh's community council: In the summer 1945 an American occupation officer called on the Mayor and enquired, on behalf of Einstein, what had become of his house and his sailboat. Regarding the sailboat's subsequent owner, he was given the wrong information. That is how Einstein came to believe that a Nazi had got it. Since then, I was able to clarify the matter. The buyer of the boat, a dentist in Babelsberg, was not a Nazi and had nothing to do with the Nazis. On the contrary! He supported the orphans of an antifascist who had been condemned to death by a "Peoples' Court" [*Volksgerichtshof*] and had been executed. It is regrettable that it was too late for Einstein to learn the true state of affairs. He had been particularly attached to his sailboat, not just because it was so comfortable but also because of the many glorious hours he experienced on it with his friends.

HW: Herr Professor loved his sailboat very much. That was already the case in Caputh and is also shown by his letter to me.

FH: You told me that the Einsteins sent you several food parcels.

HW: The first one came just a few days after the arrival of the air mail letter of April 8, 1947.

As I recall, a parcel arrived every six to eight weeks, and there may have been ten parcels, altogether. After the first one, I wrote that, much as I loved coffee, I had traded some of the coffee beans for fruit for my boy, and so they sent me twice the amount the next time.

FH: When you look back today to the time you spent in Einstein's household, would you have missed those six years of your life?

HW: Absolutely not.

FH: Let us then leave the questions and answers at that. By continuing our conversation this or that recollection might come to light, but we are not striving for completeness. You have, indeed, had many diverse experiences with Einstein and his family, and aided by your good memory you have recounted them objectively and truthfully. While the possibility of error cannot be excluded in dealing with events from the distant past, nothing presented here was invented or "massaged"—in contradistinction to some biographical accounts about Einstein, that abound with fantasies and legends.

When I occasionally asked you about minute matters and inquired about seemingly unimportant details, it was in order to portray the environment of the great physicist and humanist during his last years in Berlin as factually and candidly as possible, and to capture something of the atmosphere that permeated his home. In doing so, we spoke openly about matters that are otherwise gladly circumvented, but I believe that it accords with Einstein's sentiment to be candid and outspoken here, as well. In 1950, when Max von Laue drew Einstein's attention to some critical comments made by his biographer Philipp Frank, he replied, "That is just the way Dear God created his supposedly favorite creature." Einstein did not wish to appear as anything other than what he was.

All that remains for me to do is to thank you very warmly, esteemed Frau Waldow, for answering my exhaustive questions so patiently and for giving me, now and then, a cue to provide supplementary biographical information and to comment on the history of science.

HW: I was glad to do so, Herr Professor . . .

FH: . . . wherein "Herr Professor" is on this occasion not synonymous with *Albert Einstein*.

NOTES

CHAPTER 1: INTRODUCTORY REMARKS

1. Friedrich Herneck, *Einstein privat. Herta W. erinnert sich an die Jahre 1927 bis 1933* (Berlin: Buchverlag Der Morgen, 1978).

2. D. Hoffmann and D. B. Herrmann, "Friedrich Herneck zum Gedenken," *NTM International Journal of History and Ethics of Natural Sciences, Technology and Medicine* 2, no. 1 (1994): 183–84.

3. Dieter B. Herrmann, "Friedrich Herneck als Hochschullehrer an der Humboldt-Universität zu Berlin," in *Friedrich Herneck: Ein Leben in Suche nach Wahrheit*, ed. A. Wessel, D. B. Herrmann, and K. F. Wessel (Berlin: Logos-Verlag, 2016), p. 110.

4. Friedrich Herneck, *Albert Einstein. Ein Leben für Wahrheit, Menschlichkeit und Frieden* (Berlin: Buchverlag der Morgen, 1963), p. 7.

CHAPTER 2: FRIEDRICH HERNECK, HISTORIAN OF SCIENCE IN DIFFICULT TIMES

1. Dieter B. Herrmann, "Friedrich Herneck als Hochschullehrer an der Humboldt-Universität zu Berlin," in *Friedrich Herneck: Ein Leben in Suche nach Wahrheit*, ed. A. Wessel, D. B. Herrmann, and K. F. Wessel (Berlin: Logos-Verlag, 2016), p. 109.

2. Albert Einstein, "Why Socialism?" *Monthly Review* 1, no. 1 (May 1949).

3. An extensive publication dealing with the life and work of Friedrich Herneck is due to appear shortly: Karl-Friedrich Wessel, Dieter B. Herrmann, and Andreas Wessel, eds. *Friedrich Herneck—ein Leben in Suche nach Wahrheit* (Berlin: Logos Verlag, 2016), in the series: Berliner Studien zur Wissenschaftsphilosophie und Humanontogenetik, Vol. 32.

CHAPTER 3: EINSTEIN'S ROAD TO BERLIN—AND BEYOND

1. Wilhelm Heinrich Solf's (1862–1936) report of January 3, 1922, appears in: Siegfried Grundmann, *Einsteins Akte: Wissenschaft und Politik—Einsteins Berliner Zeit* (Berlin: Springer, 2004), pp. 231–35.

2. Max Planck (1858–1947) was a distinguished theoretical physicist, best known for introducing the concept of quantization into physics and discovering the fundamental constant *h* that bears his name. By postulating that a radiating body had quantized (i.e., discrete) energy levels, he was able to explain the spectrum of "black body" radiation (1900); but he initially resisted Einstein's generalization of that idea, when Einstein postulated the existence of light quanta (photons) in his explanation of the photoelectric effect (1905). Planck was appointed permanent secretary of the Kaiser Wilhelm Society (now Max Planck Society) in 1912 and was instrumental, along with Walther Nernst, in bringing Einstein to Berlin and to the Prussian Academy of Sciences.

An excellent pianist, also an organist and cellist, Planck shared an enthusiasm for chamber music with Einstein, and the two often played together. As a keen patriot, he was one of the signatories of the Manifesto of the Ninety-Three in defense of Germany's military in 1914, though he later backed away from it.

Planck's personal life was marked by tragedy of epic proportions. His first wife, with whom he had two boys and twin daughters, died in 1909. One of his sons was killed in WWI, and both his daughters died in childbirth. His home and all his possessions were destroyed in a bombing raid in 1944 and, in the same year, his surviving son was implicated in the conspiracy to assassinate Hitler. Even though Planck implored Himmler, Göring, and Hitler to save his son's life, Erwin Planck was executed in 1945. Planck died in 1947 while staying with relatives in Göttingen. (Fritz Stern, *Einstein's German World* [Princeton: Princeton University Press, 1999], pp. 35–58.)

3. For example: R. Highfield and P. Carter, *The Private Lives of Albert Einstein* (London: Faber & Faber, 1993); Walter Isaacson, *Einstein, His Life and Universe* (New York: Simon & Schuster, 2007); Albrecht Fölsing, *Albert Einstein: A Biography*, trans. E. Osers (New York: Penguin Books USA, 1997).

4. Walther Herrman Nernst (1864–1941) was a physicist who won the Nobel Prize in Chemistry in 1920. Along with Max Planck, he was instrumental in bringing Einstein to Berlin. He was a vocal critic of Hitler and Nazism. His two sons died in WWI, and his three daughters all married Jewish men.

5. Arnold Sommerfeld (1868–1951) was a German physicist who made important contributions to atomic physics and quantum mechanics. He received numerous honors, including the prestigious Planck medal, and has the distinction of having nurtured more Nobel Prize winners as students than anyone else.

Max Born (1882–1970) was a German Jewish physicist who made fundamental contributions to quantum mechanics in the 1920s, for which he was awarded the Nobel Prize in physics as late as 1954. Dismissed from his post in Göttingen in 1933, he taught at the University of Edinburgh until his retirement. He returned to Germany in 1954 and died in Göttingen.

Following their 1909 meeting in Salzburg, Einstein became a life-long friend of Born and Born's wife, Hedwig, and they began a voluminous correspondence that lasted over forty years.

The Born-Einstein Letters. The Correspondence between Albert Einstein and Max and Hedwig Born 1920–1955, commentaries by Max Born, trans. I. Born (New York: Walker & Co., 1971).

6. Fritz Haber (1868–1934) was a brilliant physical chemist best known for discovering a process by which nitrogen in the air is "fixed" and ammonia is produced, for which he received the Nobel Prize in 1918. The process is of great economic importance for the production of fertilizers and explosives. During World War I, Haber played a leading role in the development of chemical weapons and supervised the first deployment of poison gas on the Western front. He was a close friend of Einstein, although their political views differed radically. Haber, a Jew, believed strongly in assimilation, while Einstein thought that Zionism offered the best hope for Europe's Jews. Soon after Hitler gained power, Haber resigned from the Kaiser Wilhelm Institute that he headed, although he could have remained on account of his war service. He died shortly afterward in Switzerland. See: Dietrich Stoltenberg, *Fritz Haber: Chemist, Nobel Laureate, German, Jew* (Philadelphia: Chemical Heritage Foundation, 2005), pp. 121–70. Also: Stern, *Einstein's German World*, pp. 59–164.

7. For more about Archenhold and his fate following Hitler's accession to power, see chapter 4, Second Conversation. Another of Einstein's popular lectures on relativity is mentioned in chapter 4, Fourth Conversation.

Dieter B. Herrmann, "Einstein and Archenhold: Two Champions for the Popularization of the Natural Sciences," in *Engineer of the Universe. Albert Einstein. Hundred Authors for Einstein*, ed. Jürgen Renn (Weinheim: Wiley-Verlag, 2005), pp. 234–37.

8. Sir Arthur Eddington (1882–1944) was an eminent astronomer, physicist, and mathematician, who shared more than scientific interests with Einstein: He was

a Quaker and conscientious objector during WWI, and after it ended he worked with Einstein toward re-establishing contacts between German and Allied scientists.

9. Harry Graf Kessler (1868–1937) served as a German diplomat under Wilhelm II and became a stout supporter of the Weimar Republic after the war. He was a friend and political ally of Walther Rathenau and Einstein, a patron of the arts, and a prolific and perceptive diarist.

Harry Kessler, *Tagebücher 1918–1937* (Berlin: Deutsche Buchgemeinschaft, 1967), pp. 278–80.

10. Hermann Anschütz-Kämpfe (1872–1931) was the son of a mathematics teacher. Following his father's death, he was adopted by the Austrian art historian Kämpfe. Anschütz studied art history and became an experienced, independently wealthy explorer. He became interested in the gyrocompass as a navigation instrument while making plans to reach the North Pole by submarine—magnetic compasses being useless in the vicinity of the magnetic pole. Einstein worked closely with engineers at Anschütz's company on detailed designs that improved the precision and reliability of gyrocompasses, the standard navigational device on ships and planes before the advent of global satellite positional technology.

The living room of Einstein's ground floor apartment in Kiel opened onto a lawn that led down to a pier where his sailboat, the *Lisa*, was tied up. For particulars and many letters between Anschütz, Einstein, and also Elsa, see: Dieter Lohmeier and Bernhardt Schell, *Einstein, Anschütz und der Kieler Kreiselkompass* (Heide in Holstein: Verlag Boyens, 1992); see also: Fölsing, *Albert Einstein: A Biography*, pp. 400–402.

11. For a comprehensive account of Einstein's far-flung sea voyages (1922–1933) that is based on his travel diaries, see: Josef Eisinger, *Einstein on the Road* (Amherst: Prometheus Books, 2011).

12. János Plesch (1878–1957), born in Budapest to a family of physicians, trained in several medical specialties before settling in Berlin in 1903. He was a medical researcher, as well as serving as a physician with a large private practice, with many prominent patients among them. He was at the center of Berlin's social scene and was the owner of the "Villa Lemm," the family's country home in Gatow near Berlin. He put the Gatow estate at Einstein's disposal to serve as a hideaway. Plesch emigrated to England in 1933 and worked as pathologist at St. George Hospital in London. See: John Plesch, *János, the Story of a Doctor* (New York: A. A. Wyn, 1949).

13. Henry Goldman (1857–1937) was instrumental in creating the financial conglomerate of Goldman Sachs in the early twentieth century. He was a stalwart supporter of Germany before and during WWI, and in 1915 he broke with his partner Sachs, who supported the Allies' cause, and left the company. In the course of his fre-

quent visits to Germany, Goldman, who spoke German fluently, met and befriended Albert Einstein in 1924, and through him met the physicist Max Born, who was then doing seminal work in quantum mechanics in Göttingen. Goldman generously donated funds for research to Born's department, as well as to Otto Stern and Walther Gerlach, who performed their celebrated atomic beam experiment that demonstrated that an atom's electrons carry "spin." After Hitler came to power, Goldman witnessed the disastrous changes in Germany personally and helped many Jewish intellectuals to escape.

June Breton Fisher, *When Money was in Fashion: Henry Goldman, Goldman Sachs, and the Founding of Wall Street* (New York: Palgrave Macmillan, 2010), p. 144.

14. Erwin Schrödinger (1887–1961) and his wife were frequent sailing companions of Einstein. Schrödinger was the discoverer of wave mechanics, now referred to as quantum mechanics, for which he was awarded the Nobel Prize in 1933. Schrödinger was also interested in philosophy, and with his book *What Is Life?* he aroused the interest of many physicists in molecular biology.

Although Schrödinger was not Jewish, he had no wish to live in Berlin under the Nazis, and obtained a position at an Oxford college in 1933 with the help of Frederick Lindemann, the director of Oxford's Clarendon laboratory. Lindemann, a friend of Churchill and of Einstein, was instrumental in helping a number of German Jewish scientists make their escape to England. After a short stay in Oxford, Schrödinger accepted a professorship in Graz, Austria—not without misgivings, for the University was a known hotbed of Nazi activity. Following Hitler's annexation of Austria in 1938, Schrödinger was obliged to publish in newspapers an abject confession of past political errors and an appeal to Austrian scientists to work in accord with the will of the Führer. He was dismissed all the same, as being politically unreliable. (The University of Berlin even stripped him of his Professor Emeritus title!) Disguised as weekend tourists, he and his wife escaped from Austria by car into Italy and returned to Oxford. Schrödinger spent the war years at the Institute for Advanced Studies in Dublin, where his penchant for a *ménage á trois* caused a mild sensation.

Walter Moore, *Schrödinger: Life and Thought* (Cambridge: Cambridge University Press, 1989), pp. 320–414.

15. The conversation between Einstein and Rabindranath Tagore was arranged by Bruno Mendel, Toni's son-in-law, and it took place in Toni's Wannsee villa in July 1930. Mendel was a medical researcher whose private laboratory was on the grounds of the villa. Einstein often visited him there to observe his experiments and to offer suggestions. (See chapter 4, Fifth Conversation.) *Das goldene Boot,* ed. M. Kämpfchen, trans. A. O. Carius (Düsseldorf: Artemis & Winkler, 2005). pp. 547–55.

16. Albert Abraham Michelson (1852–1931) graduated from the US Naval Academy and conducted his first measurements of the velocity of light in 1877 in Annapolis. After resigning from the Navy, he performed the famous Michelson-Morley experiment at the Case Institute in Cleveland in 1887. He used an interferometer of his own design to demonstrate that the velocity of light is the same in all directions, thereby ruling out the existence of an ether. Michelson was the first American to be awarded a Nobel Prize in science.

17. For an account of Einstein's musical activities in Oxford, see: Eisinger, *Einstein on the Road*, pp. 126–31.

18. Max von Laue (1879–1960) investigated the properties of X-rays following their discovery by Wilhelm Röntgen in 1895. He demonstrated their wavelike character by employing a crystal as a natural diffraction grating and used the diffraction pattern to determine the arrangements of the atoms in the crystal. Einstein considered this work the most beautiful discovery in physics. X-ray diffraction analysis has since become a powerful investigative tool in condensed matter physics, as well as in molecular biology. Von Laue, an early champion of the relativity theory, was awarded the Nobel Prize in physics in 1917.

Von Laue was the only prominent physicist to protest the expulsion of Einstein from the Academy, and he remained in contact with his former Jewish colleagues after the Nazis came to power. He was also the only German physicist with whom Einstein corresponded after the Second World War. For a concise, fascinating account of how science and scientists fared in the Third Reich, see: Walter Gratzer, *The Undergrowth of Science: Delusion, Self-Deception and Human Frailty* (Oxford: Oxford University Press, 2000), pp. 219–80.

19. Erwin Finlay Freundlich (1885–1964) was the first astronomer intent on measuring the effects of the gravitational field on light that are predicted by the general relativity theory. He mounted an expedition to Crimea to observe the deflection of starlight by the sun's gravitational field during a total eclipse in 1914, but when the First World War broke out, he and his party were interned by Russia before they were able to perform the experiment. He had a Jewish wife and a Jewish grandmother and was therefore obliged to emigrate in 1933. He went first to Istanbul, then to Charles University in Prague, from where he had to leave again when Hitler occupied Czechoslovakia, and finally to St. Andrew's University in Scotland. After his retirement he returned to his native Germany.

For a detailed account of the architecture and science of the Einstein Tower, as well as Freundlich's troubled relations with Einstein and with Ludendorff, see: Klaus Hentschel, *The Einstein Tower*, trans. A. M. Hentschel (Stanford: Stanford University Press, 1997).

Hans Ludendorff was the brother of General Erich Ludendorff, the joint

Commander-in-Chief (with Paul von Hindenburg) of the German army during WWI. Erich Ludendorff was a devoted nationalist who had a brief flirtation with Hitler that he later came to regret. After Hindenburg, as president, turned the government over to Hitler in 1933, Ludendorff charged him with having delivered the fatherland to "the greatest demagogue of all time . . . [and that] this accursed man will cast our Reich into the abyss and bring our nation to inconceivable misery. Future generations will damn you in your grave for what you have done." (Ian Kershaw, *Hitler 1889–1936: Hubris* [New York: Norton, 1998], p. 427.)

20. For more about Joseph and Boris Schwarz, see: chapter 4, note 25.

21. Fisher, *When Money Was in Fashion*, pp. 136–50.

22. Niels Bohr (1885–1962) was a Danish theoretical physicist who in 1920 proposed the first crude model of the atom with quantized energy levels, for which he was awarded a Nobel Prize in 1922: The model has the electrons surrounding the nucleus confined to discrete energy levels (orbits), and they can make transitions between them by absorbing or emitting photons. The initial "Bohr model" of the atom was refined with the advent of quantum mechanics and the discovery of electron "spin" and was successful in explaining the chemical properties of the elements.

Bohr met Einstein for the first time when he visited Berlin in 1920, and their letters show that each of them greatly admired the other. Their difference with regard to the probabilistic interpretation of quantum mechanics did not come to a head until they met at the 1927 Solvay Conference in Brussels, and it continued to rankle from then on. For a detailed account of their debate, see: Abraham Pais, *Niels Bohr's Times in Physics, Philosophy and Polity* (Oxford: Clarendon Press, 1991), pp. 295–323.

23. Isaacson, *Einstein: His Life and Universe*, p. 298.

24. Queen Elisabeth was a violinist and used to play trios with Einstein and a lady-in-waiting. She was a member of the royal Wittelsbach family and the niece of the ill-fated Elisabeth of Austria, the wife of Emperor Franz Josef. The Wittelsbach dynasty ruled Bavaria from 1180 to 1918.

25. For additional details about Locker-Lampson and Einstein's activities in Britain, see: Isaacson, *Einstein: His Life and Universe*, pp. 419–24.

26. Alice Calaprice has pointed out that the household at 112 Mercer Street now consisted of Einstein, Maja, Margot, Helen Dukas, a dog, Chico, a cat, Tiger, and a parrot, Bibo. All of them continued to live in that house until their deaths, Margot being the last to die—in 1986. (Alice Calaprice, *The Einstein Almanac* [Baltimore: Johns Hopkins Press, 2005], p. 118.)

27. Albert Einstein to Toni Mendel, March 24, 1948, Princeton, New Jersey, in private possession.

28. Max Born, *The Born-Einstein Letters 1916–1955: Friendship, Politics and Physics in Uncertain Times* (New York: MacMillan, 2005), p. 229. Also: Herneck, *Albert Einstein*, p. 109.

CHAPTER 4: EINSTEIN AT HOME

1. Philipp Frank, *Einstein: His Life and Times* (New York: Alfred Knopf, 1947), p.124. Philipp Frank (1884–1966) was a theoretical physicist and philosopher who had a closer personal relationship with Einstein, and knew him better than most of his colleagues. After describing the Haberlandstrasse apartment, Frank added this insightful comment: ". . . but when one entered this home, one found that Einstein still remained a 'stranger' in such a surrounding—a bohemian in a middle-class home."

2. In Germany and Austria the term is used mockingly to characterize a conventional middle-class mentality.

3. John Plesch, *János: The Story of a Doctor* (New York: A. A. Wyn, 1949), p. 217. See also: chapter 3, note 12.

4. See chapter 3, note 2.

5. Rudolf Kayser, *Stendhal, oder das Leben eines Egotisten* (Berlin: S. Fischer Verlag, 1928); in English: Rudolf Kayser, *Stendhal: The Life of an Egotist* (New York: Henry Holt, 1930).

6. Plesch, *János: The Story of a Doctor*, p. 209.

7. Plesch recalls that at his country estate in Gatow, Einstein ate between five and ten pounds of strawberries at one sitting, on more than one occasion. Plesch, *János: the Story of a Doctor*, p. 206.

8. The physicist Max Born was one of Einstein's oldest friends. See chapter 3, note 5.

9. See chapter 3, note 13. June Breton Fisher, *When Money Was in Fashion: Henry Goldman, Goldman-Sachs, and the Founding of Wall Street* (New York: Palgrave Macmillan, 2010), pp. 137–50.

10. This description fits the Silex vacuum coffee brewer of that period, although coffeemakers based on the same design principle go back to the 1830s.

11. See chapter 3, notes 2 and 18.

12. Friedrich Simon Archenhold (1861–1939) was a German Jewish astronomer who founded the Treptow Observatory in a suburb of Berlin. It houses the world's

longest, movable refracting telescope, with a focal length of twenty-one meters (sixty-eight feet). Archenhold believed in popularizing science, and in 1915 Einstein gave his first public lecture on the theory of relativity at the Treptow Observatory, which now bears Archenhold's name. After Hitler gained power, Archenhold was forced to give up his position. He died in Berlin, but his wife, who was his collaborator, and his daughter, perished in the Theresienstadt (Terezin) concentration camp.

See chapter 3, note 8. Also: Dieter B. Herrmann, "Einstein and Archenhold: Two Champions for the Popularization of the Natural Sciences," in *Albert Einstein: Chief Engineer of the Universe: One Hundred Authors for Einstein*, ed. Jürgen Renn (Weinheim: Wiley-Verlag, 2005), pp. 234–37.

13. Plesch, *János: The Story of a Doctor*, p. 209.

14. Plesch, *János: The Story of a Doctor*, p. 206.

15. See chapter 3, note 2.

16. See chapter 3, note 18.

17. See chapter 3, note 12.

18. Cart Seelig, *Helle Zeit—Dunkle Zeit* (Berlin: Springer Vieweg, 1956), pp. 45–48.

19. See chapter 3, note 4.

20. For more about Haber, see chapter 3, note 6.

21. Heinrich Mann (1871–1950) was the first son of a Hanseatic merchant family and the brother of the more celebrated Thomas. He is the author of *Professor Unrat*, made famous by Sternberg's film *The Blue Angel*. When WWI broke out, he protested against the general war hysteria in his essay *Zola*, and against the position taken by his younger brother Thomas. This led to a break between the two that was not healed for eight years. Heinrich Mann was a supporter of the Weimar Republic and together with Einstein he signed an appeal to the social democratic and communist parties to unite in their stand against the Nazis. With Hitler coming to power, he emigrated to France, where he participated in antifascist activities before fleeing to the United States in 1940. There, he worked as a script writer for the Warner Brothers studio. He was honored by the government of the DDR in 1949, but he died in Santa Monica before he could move to Berlin.

22. Thomas Mann (1875–1955) was the second son of an old Lübeck merchants' family, which is portrayed in his best-known prose work, *The Buddenbrooks*, for which he received the Nobel Prize in 1929. In 1914 he was one of the signers of the infamous Manifesto of the Ninety-Three in support of the German military. He took a political stand against the Nazis and was on a European journey when they came to power. He did not return to Germany and emigrated to the United States in 1938. During

the war he gave monthly talks that were broadcast to Germany by the BBC. He became a US citizen in 1940 and, after being accused of being a "fellow traveler" of communists, he moved to Zurich where he died. His many works include *The Magic Mountain*, *Joseph and his Brothers*, *Doctor Faustus*, and *Death in Venice*.

23. Gerhart Hauptmann (1862–1946) was a German dramatist and novelist who was awarded the Nobel Prize for literature in 1912. His writings had an antimilitaristic flavor, and Kaiser Wilhelm II strongly disapproved of him. He was, nonetheless, among the artists, intellectuals, and scientists who signed the Manifesto of the Ninety-Three in support of Germany's role in 1914. His best known play, *The Weavers*, deals with the working condition of Silesian weavers and their insurrection in 1840. Though he had applied for membership in the Nazi party, he was not trusted by Goebbels, and his writings were censored or suppressed.

24. Hedwig Wangel (1875–1961) interrupted her successful stage career in 1909 to devote herself to the rehabilitation of women who had served time in prison. She resumed acting in 1924 and directed two films in support of her social work. She also played supporting roles in numerous other films. In 1941 she played the role of a duplicitous Queen Victoria in the widely seen, anti-British, Nazi propaganda film *Ohm Krüger*, which deals with the Boer war.

25. Joseph Schwarz was a concert pianist who, during the 1920s, came to the Haberlandstrasse together with his violinist son, Boris, "once or twice a month" to play piano trios with Einstein. Boris recalls how each session was followed by Frau Elsa's "coffee and cake." Einstein took a lively interest in Boris's career, and in 1936 he sent him an affidavit to allow him to come to New York. With Einstein's assistance, Boris established himself sufficiently as violinist to bring his parents to the United States as well. Occasionally Boris still played with Einstein in Princeton. See: Boris Schwarz, "Musical and Personal Reminiscences of Albert Einstein," in *Albert Einstein: Historical and Cultural Perspectives*, ed. G. Holton and Y. Elkana (Mineola, New York: Dover Publications, 1979), pp. 409–16.

26. See chapter 3, note 15. Rabindranath Tagore (1861–1941) was a renowned Bengali writer, poet, musician, and graphic artist, with an enthusiastic following in Europe in the aftermath of the First World War. Tagore and Einstein were widely perceived as representatives of Eastern and Western cultures, respectively, and two of their dialogues have been preserved and published, the first taking place in the Haberlandstrasse apartment (1926), and the other in Toni Mendel's Wannsee villa (1930). *Das goldene Boot*, ed. N. Kämpfchen. pp. 547–55.

27. Essen is an industrial town in the German Ruhr region; and "Essen" means "edibles" in German.

28. Albert Einstein, *Mein Weltbild*, ed. C. Seelig (Zurich: Europa Verlag, 1934), p. 50.

29. The Einsteins had rented a house in Scharbeutz, a small seaside town on the Bay of Lübeck.

30. Hermann Struck (1876–1944) made a superb lithograph of Einstein on that occasion. See: Josef Eisinger, *Einstein on the Road* (Amherst, NY: Prometheus Books, 2011), p. 63.

31. See chapter 3, note 18.

32. Plesch, *János: The Story of a Doctor*, p. 207. ". . . later on" refers to Einstein's uncompromising rejection of almost all ties to Germany, after the Second World War.

33. Friedrich Herneck, *Einstein und sein Weltbild* (Berlin: Buchverlag Der Morgen, 1976); also: Friedrich Herneck, *Bahnbrecher des Atomzeitalters* (Berlin: Buchverlag Der Morgen, 1980).

34. Erich Mühsam (1878–1934) was a well-known writer, poet, and anarchistic political activist, who took part in the formation of the short-lived socialist Räterepublik in Bavaria (1919). He served several prison sentences and was among the first to warn the divided Left about the Nazis. He was arrested in 1933, he was tortured, and nine months later he was murdered by the SS in the Sachsenhausen concentration camp.

35. On August 24, 1920, a rally in opposition to the relativity theory took place in the large hall of the Berliner Philharmonie. The rally was organized by Paul Weyland, an anti-Semitic activist who had a long career as a charlatan and confidence artist. He was also responsible for staging the famous debate between Einstein and the physicist Philipp Lenard during a scientific meeting in Bad Nauheim. Weyland survived the war and came to the United States, where he accused Einstein of being a communist. For an account of Weyland's strange life story by Andreas Kleinert, see: http://www.physik.uni-halle.de/Fachgruppen/history/weyland.htm.

36. Josef Popper-Lynkeus (1838–1921) was a liberal, democratic social philosopher, scientist, and prolific author.

37. Rudolf Hilferding (1877–1941) was a Jewish, Austrian-born Marxist theoretician and economist who wrote the seminal financial treatise *Das Finanzkapital* in 1910. He twice served as economics minister of the Weimar Republic. Following the Nazi occupation of Austria he fled to Zurich, and later to Paris, where he was arrested and tortured, before being murdered by the Gestapo.

38. Max Born, *The Born-Einstein Letters, 1916–1955: Friendship, Politics and Physics in Uncertain Times* (New York: Macmillan, 2005), pp.125–27.

39. Friedrich Herneck, *Albert Einstein* (Berlin: Buchverlag Der Morgen, 1963), pp. 13–14.

40. Albert Einstein, *Gelegentliches von Albert Einstein* (Berlin: Aldus Druck, 1929).

41. Friedrich Herneck, "Albert Einstein und das politische Schicksal seines Sommerhauses in Caputh bei Potsdam," in Herneck, *Einstein und sein Weltbild*, pp. 256–73.

42. The carriage way is now named *Am Waldrand*.

43. Friedrich Herneck, *Albert Einstein—Ein Leben für Wahrheit, Menschlichkeit und Frieden* (Berlin: Buchverlag Der Morgen, 1963).

44. The reference is to the Swedish film *She Danced for Only One Summer*, which is based on Per Olof Ekström's novel *Sommardansen*.

45. Herneck's high praise for Walter Friedrich's musicianship, his accomplishments, and his views should be seen in light of Friedrich's considerable political clout at Herneck's university.

Many professional musicians were impressed by Einstein's musicianship, if not by his bowing technique, and enjoyed playing with him. For a more comprehensive assessment of Einstein's violin playing, see: Peregrine White, "Albert Einstein: The Violinist," *The Physics Teacher*, 43 (May 2005): 286–88. (To download, go to: astro1.panet.utoledo.edu/~ljc/Ein_violin.pdf). See also: Eisinger, *Einstein on the Road*, pp. 126–31.

Herneck is somewhat dismissive of Plesch's book, which indeed contains factual errors, but Plesch's account of Einstein's personality is remarkably perceptive: Plesch, *János: The Story of a Doctor*, pp. 200–18.

46. Helene Weigel (1900–1971) was a well-known Austrian-born actress and served as the artistic director of the Berliner Ensemble after Brecht's death. She was the second wife of the playwright and director Bertholt Brecht (1898–1956), the librettist of the *Three Penny Opera*. Weigel created the role of *The Mother* in Brecht's play of that name. After WWII Weigel and Brecht lived and worked in the DDR.

47. The Sanssouci Palace in Potsdam is the former summer palace of Frederick the Great. It is counted among the German rivals of Versailles.

48. See chapter 3, note 22.

49. Herneck, *Einstein und sein Weltbild*; Friedrich Herneck, *Bahnbrecher des Atomzeitalters* (Berlin: Buchverlag Der Morgen, 1980); Friedrich Herneck, *Albert Einstein* (Leipzig: B. G. Teubner Verlagsgesellschaft, 1986).

50. For Michelson, see chapter 3, note 16. Gustav Stresemann (1878–1929) was a German politician who held several ministerial posts in the Weimar Republic and was briefly Chancellor.

51. Herneck's two transcriptions are found in:

Friedrich Herneck, "An Unjustly Forgotten Speech of Albert Einstein,"

Naturwissenschaften 48, no. 2 (1961): 33. In a speech opening a broadcasting exhibition in 1930, Einstein reminds his audience, "present and absent," of the scientists that made radio possible (Oersted, Bell, Maxwell, Hertz); he exhorts them not to utilize the wonders created by science and technology with as little intellectual curiosity as the cow has in the botany of the plants it enjoys.

Friedrich Herneck, "Albert Einstein's Spoken Credo," *Naturwissenschaften* 53, no. 8 (1966): 198. On this record Einstein says how fortunate he was to be working in science and to be largely independent of the actions of others; that this independence must not blind one to his duties toward others, that he had never strived for comfort and luxury, that he even despises it a little, that he never sought rank or property, and that social justice was his passion.

52. Einstein, *Gelegentliches von Albert Einstein*.

53. Käthe Kollwitz (1867–1945) was a German painter, sculptor, and printmaker whose work often depicted the miseries of war and poverty. Ernst Toller (1893–1939) was a German expressionist playwright with Leftist leanings.

54. Martin Andersen Nexø (1869–1954) was a Danish writer who depicted the working class. Henri Barbusse (1873–1935) was a French novelist. Both were members of the Communist Party.

55. For additional details about Schrödinger and the Nazis, see chapter 3, note 14.

56. Lise Meitner (1878–1968) was a Jewish Austrian-born physicist, best known as the co-discoverer—with Otto Hahn and Fritz Strassmann—of nuclear fission. As an Austrian national she was able to continue working in Berlin until 1938, when Hitler annexed Austria and she became a German citizen. Meitner fled to Sweden, where she informed her nephew and fellow physicist Otto Frisch of Otto Hahn and Fritz Strassmann's recent experiments on the bombardment of uranium by neutrons. She and Frisch coined the word "fission" for the nucleus splitting into two nuclei of similar size and pointed out that an enormous amount of energy is released in the process. This is the work that prompted Einstein to write to President Roosevelt, alerting him to the devastating military potential of that discovery.

57. Actually, Toni Mendel emigrated to Canada, as did Bruno and Hertha Mendel and their three children—all of whom had shared her spacious Wannsee home in their years in Berlin. Toni lived in Oakville, Ontario, and maintained a sporadic correspondence with Einstein in Princeton. She died in Oakville in 1956.

58. The eminent zoologist and philosopher Ernst Haeckel corresponded with the much younger autodidact Frida von Uslar-Gleichen during his most creative years. Their love affair is the basis of the novel *Franziska von Altenhausen* by Johannes Werner.

59. Otto Heinrich Warburg (1883–1970) was an MD and a notable biochemical researcher who won the Nobel Prize in physiology in 1931. Although he had a Jewish father, he was allowed to continue working under the Nazis and was exempted from the strictures of the racial laws.

60. Albert Einstein, *Über den Frieden: Weltordnung oder Weltuntergang?* (Bern: Verlag Peter Lang, 1975)

61. Paul Dessau (1894–1979) was a German Jewish composer and conductor who emigrated to France in 1933 and to the United States in 1939. He worked in Hollywood and moved to East Germany in 1947. His opera "Einstein" was produced in Berlin in 1974.

62. During the *Kristallnacht* (night of broken glass), on November 9. 1938, organized Nazi mobs torched hundreds of synagogues and destroyed thousands of Jewish shops. They also arrested tens of thousands of Jews and shipped them to concentration camps. These synchronized attacks took place in communities all over Germany and the recently annexed Austria.

Michail Romm (1901–1971) was an eminent film director, whose *Ordinary Fascism* (1965) portrayed the rise of the Nazis in Germany. In the United States, the film was released as *Triumph over Violence*.

63. Rudolf Kayser, *Spinoza: Portrait of a Spiritual Hero*, trans. A. Allen and M. Newmark (New York: Philosophical Library, 1946).

64. Elsa's letter reads:

> Fräulein Herta Schiefelbein was employed in my home as housekeeper from 1 May 1927 until 1 June 1933.
>
> She is absolutely honest, reliable, and adept in all domestic arts. She is a very skillful cook.
>
> Her dismissal is the result of a move abroad. My best good wishes accompany her.
>
> Frau Elsa Einstein
>
> Coq sur Mer, May 33

65. Wilhelm Ostwald (1853–1932) was a very productive chemist and physical chemist and was awarded the Nobel Prize in 1909. Aside from his scientific work, he wrote and lectured widely on philosophical and social issues (e.g., at Harvard, MIT, Columbia in 1905/6) and played an active role in promoting his atheist and political ideas, as well as his extensive work in color theory. Ostwald took part in the peace movement before WWI but signed the Manifesto of the Ninety-Three once war had broken out. Walther Nernst was one of his students.

66. Most likely, Leopold Lichtwitz (1876–1943), who was an internist and director of the Rudolf Virchow Hospital in Berlin. After he was dismissed from his position in 1933, he emigrated to the United States and taught at Columbia University.

67. Erich Kleiber (1890–1956), born in Vienna, was the music director of the Berlin opera and conducted the first performance of Alban Berg's *Wozzek*. He was, in fact, *not* Jewish and was a staunch opponent of Nazism and Fascism.

68. This sentiment was due to Einstein having been a vocal public opponent of the Nazis while others preferred to lie low.

SELECT BIBLIOGRAPHY

Calaprice, Alice. *The Einstein Almanac.* Baltimore: Johns Hopkins Press, 2005.

Eisinger, Josef. *Einstein on the Road.* Amherst: Prometheus Books, 2011.

Fölsing, Albrecht. *Albert Einstein: A Biography.* Translated by E. Osers. New York: Viking, 1997.

Frank, Philipp. *Einstein: His Life and Times.* New York: Knopf, 1947.

Grundmann, Siegfried. *Einsteins Akte: Wissenschaft und Politik-Einsteins Berliner Zeit.* Berlin: Springer Verlag, 2004.

Highfield, Roger, and Paul Carter. *The Private Lives of Albert Einstein.* London: Faber and Faber, 1993.

Hoffmann, Dieter. *Einstein's Berlin: In the Footsteps of a Genius.* Baltimore: Johns Hopkins University Press, 2013.

Isaacson, Walter. *Einstein: His Life and Universe.* New York: Simon and Schuster, 2007.")

Kershaw, Ian. *Hitler, 1889–1936: Hubris*. New York: W. W. Norton, 1998.

Levenson, Thomas. *Einstein in Berlin*. New York: Bantam Books, 2003.

Moore, Walter. *Schrödinger: Life and Thought*. Cambridge: Cambridge University Press, 1989.

Neffe, Jürgen. *Einstein: A Biography*. Translated by S. Frisch. New York: Farrar, Straus & Giroux, 2005.

Pais, Abraham. *Einstein Lived Here*. Oxford: Clarendon Press, 1994.

———. *Niels Bohr's Times*. Oxford: Clarendon Press, 1991.

———. *"Subtle Is the Lord . . ." The Science and the Life of Albert Einstein*. Oxford: Oxford University Press, 1982.

Rowe, David E., and Robert Schulmann, eds. *Einstein on Politics: His Private Thoughts and Public Stands on Nationalism, Zionism, Peace, and the Bomb*. Princeton: Princeton University Press, 2007.

Stern, Fritz. *Einstein's German World*. Princeton: Princeton University Press, 1999.

INDEX